M208 Pu

CW00455229

The Open University

GTB3

Group actions

This publication forms part of an Open University course. Details of this and other Open University courses can be obtained from the Student Registration and Enquiry Service, The Open University, PO Box 197, Milton Keynes, MK7 6BJ, United Kingdom: tel. +44 (0)870 300 6090, e-mail general-enquiries@open.ac.uk

Alternatively, you may visit the Open University website at http://www.open.ac.uk where you can learn more about the wide range of courses and packs offered at all levels by The Open University.

To purchase a selection of Open University course materials, visit http://www.ouw.co.uk, or contact Open University Worldwide, Michael Young Building, Walton Hall, Milton Keynes, MK7 6AA, United Kingdom, for a brochure: tel. +44 (0)1908 858793, fax +44 (0)1908 858787, e-mail ouw-customer-services@open.ac.uk

The Open University, Walton Hall, Milton Keynes, MK7 6AA.

First published 2006. Reprinted with amendments 2007.

Edited, designed and typeset by The Open University, using the Open University TeX System.

Printed and bound in the United Kingdom by Hobbs the Printers Limited, Brunel Road, Totton, Hampshire SO40 3WX.

ISBN 0 7492 0221 1

1.2

Contents

Introduction

Throughout this course we have thought of a group as a single set with an algebraic structure. But if we look back through the examples of groups that we have met, we find that often there is a second set in the background—a set on which the elements of the group are 'acting'. For example, a permutation group consists of a set of permutations, each of which acts on (permutes) a set of symbols. Similarly, a symmetry group consists of a set of symmetries, each of which acts on a figure, which is a set of points. In this final group theory unit, we explore properties of groups acting on sets.

One reason for our interest in group actions is that they give us a different way of looking at groups. We shall see how this approach helps to unify many of the ideas in the earlier group theory units.

In Section 1 we define a *group action*, and look at several examples, which recur throughout the unit. Among these examples are actions of symmetry groups, permutation groups and groups of matrices.

In Section 2 we introduce the important concepts of *orbit* and *stabiliser*, and give many examples. For finite groups, these concepts are linked by the Orbit–Stabiliser Theorem, which we discuss in Section 3. This important theorem has a number of interesting applications. In particular, we see how a theorem from Unit GTB2 relating the sizes of the kernel and image of a homomorphism with a finite domain can be easily deduced from the Orbit–Stabiliser Theorem. We also deduce an important result about the sizes of the conjugacy classes of a finite group.

We conclude, in Section 4, with a discussion of another application of the Orbit–Stabiliser Theorem, called the Counting Theorem. This theorem is applicable when we wish to count the number of possible objects of a certain type which have some degree of symmetry. An example of such a problem is to find the number of ways of colouring the faces of a cube with three colours, when two colourings are regarded as the same if a rotation of the cube takes one to the other. This is one of three counting problems solved in the video programme associated with this section.

Study guide

The sections should be studied in the natural order. However, the video programme associated with Section 4 can be watched at any time after completion of Section 1.

Section 2 is the audio section.

1   What is a group action?

After working through this section, you should be able to:

(a) explain what is meant by a *group action*;

(b) construct an action table for a given finite group action;

(c) check the axioms for a group action.

1.1 Introduction

Before giving the general definition of a group action, we illustrate some of the ideas involved by looking at a few examples.

Consider first the group S_3 of all permutations of the set $\{1, 2, 3\}$. Each element of S_3 permutes the symbols 1, 2 and 3. For example,

the 3-cycle (1 3 2) maps 1 to 3, 3 to 2 and 2 to 1,

and

the 2-cycle (1 3) maps 1 to 3, 2 to 2 and 3 to 1.

We can record what each element of S_3 does to each of the three symbols 1, 2 and 3 by constructing a table, as shown below. We say that S_3 *acts on* the set $\{1, 2, 3\}$. The table below is the associated *action table* (or *group action table*). This is quite different from the group table for S_3, which records how the elements of S_3 combine with each other.

elements of
set $\{1, 2, 3\}$

$\wedge$	1	2	3
e	1	2	3
$(1\,2\,3)$	2	3	1
$(\mathbf{1\,3\,2})$	3	1	**2**
$(1\,2)$	2	1	3
$(1\,3)$	3	2	1
$(2\,3)$	1	3	2

elements of
group S_3

(1 3 2) maps 3 to 2

S_3 acting on $\{1, 2, 3\}$

To denote the effect of an element of a group on an element of a set, we adopt the following notation.

Notation   If the group element g sends the set element x to the set element y, then we write $g \wedge x = y$.

We read the action symbol $\wedge$ as 'wedge'.

For example, in S_3,

(1 3 2) sends 3 to 2, so we write $(1\,3\,2) \wedge 3 = 2$.

Exercise 1.1   Use the action table above to determine the following elements of $\{1, 2, 3\}$:

(a) $(1\,2) \wedge 2$, $(1\,2) \wedge 3$, $(1\,3\,2) \wedge 1$, $(1\,3) \wedge 2$;

(b) $e \wedge x$, for each $x \in \{1, 2, 3\}$.

We next consider how the elements of $S(\square)$, the group of symmetries of the square, act on the set of vertices of the square. Each symmetry of the square permutes the set $X = \{1, 2, 3, 4\}$; for example,

the rotation a maps vertex 1 to vertex 2, so $a \wedge 1 = 2$,

the reflection s interchanges vertices 2 and 4, so $s \wedge 2 = 4$ and $s \wedge 4 = 2$.

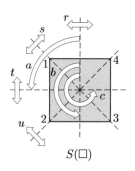

$S(\square)$

As before, we can record the effect of each symmetry of the square on each element of the set $\{1, 2, 3, 4\}$ in the form of an action table, as shown below.

$\wedge$	1	2	3	4
e	1	2	3	4
a	2	3	4	1
b	3	4	1	2
c	4	1	2	3
r	4	3	2	1
s	1	4	3	2
t	2	1	4	3
u	3	2	1	4

$S(\square)$ acting on
$X = \{1, 2, 3, 4\}$

$\circ$	e	a	b	c	r	s	t	u
e	e	a	b	c	r	s	t	u
a	a	b	c	e	s	t	u	r
b	b	c	e	a	t	u	r	s
c	c	e	a	b	u	r	s	t
r	r	u	t	s	e	c	b	a
s	s	r	u	t	a	e	c	b
t	t	s	r	u	b	a	e	c
u	u	t	s	r	c	b	a	e

$S(\square)$

From this table we can read off the effect of any symmetry on any vertex. For example,

$$a \wedge 1 = 2 \quad \text{and} \quad s \wedge 2 = 4.$$

Exercise 1.2   Use the action table for $S(\square)$ acting on $X = \{1, 2, 3, 4\}$ to do the following.

(a) Write down $r \wedge 2$, $c \wedge 3$, $b \wedge 1$ and $u \wedge 4$.

(b) Verify that

$$e \wedge x = x, \quad \text{for each } x \in X.$$

(c) Verify that

$$b \wedge (u \wedge 2) = (b \circ u) \wedge 2 \quad \text{and} \quad s \wedge (c \wedge 4) = (s \circ c) \wedge 4.$$

Can you guess a general rule which includes these two results as particular cases?

We next consider the action of $S(\square)$ on a different set. What effect does each symmetry of the square have on the set of lines of symmetry of the square?

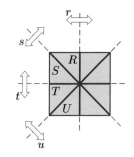

Let R, S, T and U be the parts of the axes of symmetry of the four reflections r, s, t and u, respectively, which lie within the square.

Each element of $S(\square)$ permutes the set $X = \{R, S, T, U\}$. For example, the rotation a maps the lines R and T to each other, and maps the lines S and U to each other, so

$$a \wedge R = T, \quad a \wedge S = U, \quad a \wedge T = R, \quad a \wedge U = S.$$

Also, the reflection r fixes the lines R and T, and maps the lines S and U to each other, so

$$r \wedge R = R, \quad r \wedge S = U, \quad r \wedge T = T, \quad r \wedge U = S.$$

This information enables us to complete two rows of the action table, as shown in Exercise 1.3.

Exercise 1.3   Fill in the missing entries in the following table.

∧	R	S	T	U
e				
a	T	U	R	S
b				
c				
r	R	U	T	S
s				
t				
u				

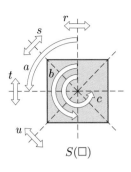

$S(\square)$ acting on
$X = \{R, S, T, U\}$

1.2 Definition of a group action

In each of the examples in the previous subsection, we have a set X and a group G that 'acts on' X in some way. More precisely, given any element g of G and any element x of X, g 'acts on' x; that is, g sends x to some element of X, which we denote by $g \wedge x$.

Now, given any set X and any group G, we can always prescribe some sort of 'action' of G on X by arbitrarily assigning values of $g \wedge x$. However, to confine ourselves to actions which prove to be of interest, we require certain conditions to hold.

In each of the examples that we have seen, the identity element e of the group $(G, \circ)$ leaves each element x of the set X *fixed*:

$$e \wedge x = x, \quad \text{for each } x \in X.$$

Moreover, in Exercise 1.2(c) we saw examples of what happens when an element x of the set is acted on by two group elements g_2 and g_1 in succession. These examples suggest that if g_2 acts on x, sending it to $g_2 \wedge x$, and then g_1 acts on the outcome, sending it on to $g_1 \wedge (g_2 \wedge x)$, the resulting element of X is the same as would be obtained by acting on x by the composite $g_1 \circ g_2$:

$$g_1 \wedge (g_2 \wedge x) = (g_1 \circ g_2) \wedge x, \quad \text{for all } g_1, g_2 \in G.$$

Acting on x by g_2 and then g_1 is the same as acting on x by $g_1 \circ g_2$.

These observations motivate our formal definition of a *group action*.

Definition   A group $(G, \circ)$ **acts on** a set X if the following three axioms hold.

GA1 CLOSURE   For each $g \in G$ and each $x \in X$, there is a unique element

$$g \wedge x \in X.$$

GA2 IDENTITY   For each $x \in X$,

$$e \wedge x = x,$$

where e is the identity element of G.

GA3 COMPOSITION   For all $g_1, g_2 \in G$ and all $x \in X$,

$$g_1 \wedge (g_2 \wedge x) = (g_1 \circ g_2) \wedge x.$$

We say that $\wedge$ is a **group action** of G on X. If $g \wedge x = y$, then we say that g **acts on** x to give y.

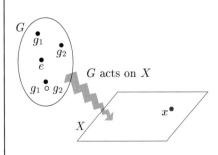

G acts on X

Remarks

1. Axiom GA1 states that each element g of the group G acts on each element x of the set X, and the result is always an element of the set X; the action does not take us out of the set X.

If it helps, you can think of $\wedge$ a a function from the set
$$\{(g,x) : g \in G, x \in X\}$$
to the set X, which maps the pair of elements (g, x) to the single element $g \wedge x \in X$.

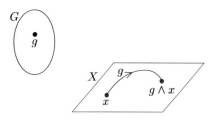

2. Axiom GA2 states that the identity element of the group *fixes* each element of the set X.

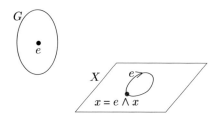

3. Axiom GA3 states that acting on x successively by two group elements g_2 and g_1 has the same effect as acting by their composite $g_1 \circ g_2$.

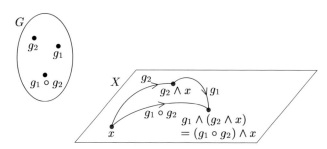

To determine whether a given operation $\wedge$ is a group action, we use the following strategy.

Strategy 1.1   To determine whether $\wedge$ is a group action.

GUESS behaviour, ... CHECK definition.

To show that $\wedge$ is a group action, show that EACH of the axioms GA1, GA2 and GA3 holds.

To show that $\wedge$ is not a group action, show that ANY ONE of the axioms GA1, GA2 or GA3 fails.

If you are unable to guess whether $\wedge$ is a group action, check each of the axioms in turn

In the next example we use Strategy 1.1 to show that the natural action of S_3 on $\{1, 2, 3\}$, which we discussed in Subsection 1.1, is a group action.

Example 1.1   Show that the operation $\wedge$ defined by
$$g \wedge x = g(x), \quad \text{for all } g \in S_3 \text{ and all } x \in \{1, 2, 3\},$$
is a group action of S_3 on $\{1, 2, 3\}$.

By the 'natural' action of a group G on a set X, we mean the action of G that sends elements of X to other elements of X in the most obvious way. (It is often possible to define other, less obvious, actions of the same group G on the same set X.)

Solution   The action table is as follows.

$\wedge$	1	2	3
e	1	2	3
$(1\ 2\ 3)$	2	3	1
$(1\ 3\ 2)$	3	1	2
$(1\ 2)$	2	1	3
$(1\ 3)$	3	2	1
$(2\ 3)$	1	3	2

We check each of the axioms GA1, GA2 and GA3 in turn.

GA1 CLOSURE   Each entry in the body of the action table is 1, 2 or 3. This confirms that the result of any element of the group S_3 acting on any element of the set $\{1, 2, 3\}$ is itself in $\{1, 2, 3\}$, so axiom GA1 holds.

GA2 IDENTITY   The first row of the action table shows that the identity element e fixes each of the three symbols 1, 2 and 3, so axiom GA2 holds.

GA3 COMPOSITION   This axiom presents more of a problem. It is tedious to use the action table to confirm that

$$g_1 \wedge (g_2 \wedge x) = (g_1 \circ g_2) \wedge x,$$

for all $g_1, g_2 \in S_3$ and all $x \in \{1, 2, 3\}$. Fortunately, there is a simpler argument available. In this example, the group elements are *functions* from the set $\{1, 2, 3\}$ to itself, and the group elements act on the elements of $\{1, 2, 3\}$ as functions; that is,

$$g \wedge x = g(x), \quad \text{for all } g \in S_3 \text{ and all } x \in \{1, 2, 3\}.$$

There are six choices for g_1, six choices for g_2 and three choices for x, making 108 combinations for g_1, g_2 and x to be checked!

So axiom GA3 is equivalent to

$$g_1(g_2(x)) = (g_1 \circ g_2)(x),$$

for all $g_1, g_2 \in S_3$ and all $x \in \{1, 2, 3\}$. This is the familiar rule for *composition of functions* and is therefore true.

Hence $\wedge$ satisfies axioms GA1, GA2 and GA3, so it is a group action.   ∎

Example 1.1 confirms that the first of the three actions introduced in Subsection 1.1 is a group action according to our definition. We show later in this subsection that the other two actions—those involving $S(\square)$—are as well.

Example 1.2   Determine whether the following table corresponds to a group action of $(\mathbb{Z}_3, +_3)$ on the set $\{a, b, c\}$.

$\wedge$	a	b	c
0	a	b	c
1	b	a	c
2	c	b	a

Solution   We check each axiom in turn.

GA1 CLOSURE   The body of the table contains only the elements a, b and c, so axiom GA1 holds.

GA2 IDENTITY   From the first row of the table, we see that 0, the identity element in $\mathbb{Z}_3$, fixes each of the elements a, b and c, so axiom GA2 holds.

GA3 COMPOSITION   A little experimentation reveals that axiom GA3 is not
satisfied. For example,

$$1 \wedge (1 \wedge c) = 1 \wedge c = c,$$

but

$$(1 +_3 1) \wedge c = 2 \wedge c = a,$$

so

$$1 \wedge (1 \wedge c) \neq (1 +_3 1) \wedge c$$

and axiom GA3 fails.

This counter-example establishes that axiom GA3 is not satisfied, so the
given table is not the table of a group action.   ∎

Exercise 1.4   Let $K = \{e, (1\ 2)(3\ 4), (1\ 3)(2\ 4), (1\ 4)(2\ 3)\}$ and let
$X = \{a, b, c, d\}$. Part of the action table of a group action $\wedge$ of K
on X is shown below. Find the entries that go in the highlighted cells,
by using axioms GA2 and GA3.

$\wedge$	a	b	c	d
e	a			
$(1\ 2)(3\ 4)$	d			
$(1\ 3)(2\ 4)$	c			
$(1\ 4)(2\ 3)$	b			

Hint: For the highlighted cells in the last three rows of the table, use
a method similar to the following. Here we find the second entry in
the second row (whose cell is not highlighted):

$$\begin{aligned}
(1\ 2)(3\ 4) \wedge b &= (1\ 2)(3\ 4) \wedge ((1\ 4)(2\ 3) \wedge a) \quad \text{(from the table)}\\
&= ((1\ 2)(3\ 4) \circ (1\ 4)(2\ 3)) \wedge a \quad \text{(axiom GA3)}\\
&= (1\ 3)(2\ 4) \wedge a \quad \text{(compose in } K)\\
&= c \quad \text{(from the table)}.
\end{aligned}$$

Inspection of an action table will quickly establish the validity, or
otherwise, of axioms GA1 and GA2. For the closure axiom GA1, we check
that every entry in the body of the table belongs to the set being acted on.
For the identity axiom GA2, we scan the row of the identity element to
check that it fixes each element of the set.

However, it is not easy to check the composition axiom GA3 from an
action table. Let us look at this problem in the context of two examples
that were introduced earlier in this section. Consider the action tables,
given below, for $S(\square)$ acting on:

(a) the set $X = \{1, 2, 3, 4\}$ of vertices of the square;

(b) the set $\{R, S, T, U\}$ of lines of symmetry of the square.

$\wedge$	1	2	3	4
e	1	2	3	4
a	2	3	4	1
b	3	4	1	2
c	4	1	2	3
r	4	3	2	1
s	1	4	3	2
t	2	1	4	3
u	3	2	1	4

$\wedge$	R	S	T	U
e	R	S	T	U
a	T	U	R	S
b	R	S	T	U
c	T	U	R	S
r	R	U	T	S
s	T	S	R	U
t	R	U	T	S
u	T	S	R	U

(a) $S(\square)$ acting on vertices

(b) $S(\square)$ acting on lines of symmetry

We can see from the tables that axioms GA1 and GA2 hold in each case, but what about axiom GA3? In fact, this axiom holds for exactly the same reason as it held in Example 1.1. In each of these two examples, the group elements are functions, the group operation is composition of functions and the group elements act on the elements of the set as functions; that is,

$$g \wedge x = g(x), \quad \text{for all } g \in S(\square) \text{ and all } x \in X,$$

where $X = \{1, 2, 3, 4\}$ for the first example and $X = \{R, S, T, U\}$ for the second example. So, in each case, axiom GA3 becomes

$$(g_1 \circ g_2)(x) = (g_1(g_2(x)), \quad \text{for all } g_1, g_2 \in S(\square) \text{ and all } x \in X.$$

This is just the rule for composition of functions, so it is true. Therefore each of these examples is a group action.

> By definition, the elements of $S(\square)$ are functions with domain and codomain $\mathbb{R}$, but we can also regard them as functions that map the set $\{1, 2, 3, 4\}$ of vertices to itself. Similarly, we can regard them as functions that map the set $\{R, S, T, U\}$ of lines of symmetry to itself.

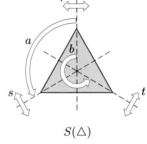

$S(\triangle)$

Exercise 1.5   For each of the following group actions, construct the action table, and verify that axioms GA1, GA2 and GA3 hold:

(a) $S(\triangle)$ acting on the set $\{1, 2, 3\}$ of vertices of the triangle;

(b) $S(\square)$ acting on the set $\{A, B, C, D\}$ of edges of the square.

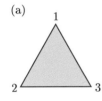

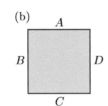

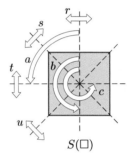

$S(\square)$

The above examples of group actions are special cases of the following result, which shows that groups of functions from a set X to itself give rise to group actions on X in a natural way.

Theorem 1.1   Let G be a group whose elements are functions from a set X to itself, with the binary operation of composition of functions. Let $\wedge$ be defined by

$$g \wedge x = g(x), \quad \text{for } g \in G \text{ and } x \in X.$$

Then $\wedge$ is a group action of G on X.

Proof   We show that axioms GA1, GA2 and GA3 hold.

GA1 CLOSURE   For each $g \in G$ and each $x \in X$, $g \wedge x$ is an element of X, since $g \wedge x = g(x)$ and g has codomain X. Thus axiom GA1 holds.

GA2 IDENTITY   The identity element e of G is the function that maps each element of X to itself, so

$$e \wedge x = e(x) = x,$$

for each $x \in X$. Thus axiom GA2 holds.

GA3 COMPOSITION   Let g_1 and g_2 be elements of G and let $x \in X$. Then

$$\begin{aligned}
g_1 \wedge (g_2 \wedge x) &= g_1 \wedge (g_2(x)) && \text{(definition of } \wedge) \\
&= g_1(g_2(x)) && \text{(definition of } \wedge) \\
&= (g_1 \circ g_2)(x) && \text{(composition of} \\
& && \text{functions)} \\
&= (g_1 \circ g_2) \wedge x && \text{(definition of } \wedge).
\end{aligned}$$

Thus axiom GA3 holds.

Hence $\wedge$ satisfies axioms GA1, GA2 and GA3, so it is a group action. ∎

Remark Theorem 1.1 is a very useful result. It immediately tells us that a whole class of tables are action tables: a natural action, such as those in Example 1.1 and Exercise 1.5, of a permutation group or a symmetry group on a set is always a group action.

Even if we are given only an action table, we may be able to interpret it as the table of a natural action of this type, and hence apply Theorem 1.1 to show that it specifies a group action. Here is an example.

Example 1.3 Let $K = \{e, (1\ 2)(3\ 4), (1\ 3)(2\ 4), (1\ 4)(2\ 3)\}$ and let $X = \{a, b, c, d\}$. Determine whether the following table specifies a group action of K on X.

$\wedge$	a	b	c	d
e	a	b	c	d
$(1\ 2)(3\ 4)$	c	b	a	d
$(1\ 3)(2\ 4)$	c	d	a	b
$(1\ 4)(2\ 3)$	a	d	c	b

Solution Although it is straightforward to check axioms GA1 and GA2 from the table, it would be a protracted task to do so for axiom GA3: we would have to check that

$$g_1 \wedge (g_2 \wedge x) = (g_1 \circ g_2) \wedge x$$

holds for all combinations of g_1, g_2 and x.

We can avoid doing any calculations and appeal directly to Theorem 1.1 if we can find some way of regarding the given table as that of a group of functions acting on the set $\{a, b, c, d\}$. Since K is the symmetry group of a rectangle (where the symmetries are represented as permutations of the vertex locations), we can ask whether the given table arises from letting the symmetries in K permute some set $\{a, b, c, d\}$ of features of the rectangle, such as the set of four vertices, or the set of four edges.

It is not difficult to see that if we label the four edges a, b, c and d, as shown in the margin, then the action table for K on $\{a, b, c, d\}$ is indeed the given one. So Theorem 1.1 confirms that this a group action. ∎

Exercise 1.6 For each of the following action tables, find a set $\{a, b, c, d\}$ of features of a rectangle such that the group

$$K = \{e, (1\ 2)(3\ 4), (1\ 3)(2\ 4), (1\ 4)(2\ 3)\}$$

acting on $\{a, b, c, d\}$ has the given action table.

(a)

$\wedge$	a	b	c	d
e	a	b	c	d
$(1\ 2)(3\ 4)$	b	a	d	c
$(1\ 3)(2\ 4)$	c	d	a	b
$(1\ 4)(2\ 3)$	d	c	b	a

(b)

$\wedge$	a	b	c	d
e	a	b	c	d
$(1\ 2)(3\ 4)$	a	d	c	b
$(1\ 3)(2\ 4)$	a	b	c	d
$(1\ 4)(2\ 3)$	a	d	c	b

1.3 Infinite actions

Until now, all our examples of group actions have involved a *finite* group G and a *finite* set X. We now turn our attention to examples in which one, or both, of the group G and the set X is infinite. For such group actions we cannot write out the entire action table, but we can sometimes present enough of it to see what is going on. Our next example illustrates this.

Example 1.4   Consider the group $(\mathbb{Z}, +)$ and the set $X = \{0, 1, 2, 3, 4, 5\}$. Show that

$$n \wedge x = (2n + x) \;(\mathrm{mod}\; 6), \quad \text{for } n \in \mathbb{Z} \text{ and } x \in X,$$

defines a group action of $\mathbb{Z}$ on X.

For example,
$$11 \wedge 4 = (22 + 4) \;(\mathrm{mod}\; 6)$$
$$= 26 \;(\mathrm{mod}\; 6)$$
$$= 2.$$

Solution   Although we cannot write out all the action table, we can present part of it, as follows.

$\wedge$	0	1	2	3	4	5
$\vdots$	$\vdots$	$\vdots$	$\vdots$	$\vdots$	$\vdots$	$\vdots$
-3	0	1	2	3	4	5
-2	2	3	4	5	0	1
-1	4	5	0	1	2	3
0	0	1	2	3	4	5
1	2	3	4	5	0	1
2	4	5	0	1	2	3
3	0	1	2	3	4	5
$\vdots$	$\vdots$	$\vdots$	$\vdots$	$\vdots$	$\vdots$	$\vdots$

The rows in the table appear to repeat for every third element of $\mathbb{Z}$.

We show that axioms GA1, GA2 and GA3 hold.

Since the group is infinite, we cannot check group action axioms GA1 and GA3 from the table, but have to use general algebraic arguments.

GA1 CLOSURE   The value modulo 6 of any integer is an element of the set $X = \{0, 1, 2, 3, 4, 5\}$, so for each $n \in \mathbb{Z}$ and each $x \in X$, we have $g \wedge x \in X$. Thus axiom GA1 holds.

GA2 IDENTITY   The identity in $\mathbb{Z}$ is 0, and we can see from the table that

$$0 \wedge x = x, \quad \text{for each } x \in X,$$

so axiom GA2 holds.

GA3 COMPOSITION   Let m and n be any elements of $\mathbb{Z}$ and let x be any element of X. We must show that

$$m \wedge (n \wedge x) = (m + n) \wedge x.$$

Now

$$m \wedge (n \wedge x) = m \wedge (2n + x) \;(\mathrm{mod}\; 6)$$
$$= (2m + (2n + x) \;(\mathrm{mod}\; 6)) \;(\mathrm{mod}\; 6)$$
$$= (2m + 2n + x) \;(\mathrm{mod}\; 6) \qquad (1.1)$$

The group operation is addition, so we write $m + n$ rather than $m \circ n$.

and

$$(m + n) \wedge x = (2(m + n) + x) \;(\mathrm{mod}\; 6)$$
$$= (2m + 2n + x) \;(\mathrm{mod}\; 6). \qquad (1.2)$$

Expressions (1.1) and (1.2) are the same, so axiom GA3 holds.

Hence $\wedge$ satisfies axioms GA1, GA2 and GA3, so it is a group action. ∎

You saw in Unit I3 that, when carrying out arithmetic modulo k, we can take the value modulo k of any integer in the calculation part-way through, or we can instead take the value modulo k at the end; we get the same answer either way. This is why we can remove the first '(mod 6)' from the second line of the calculation of $m \wedge (n \wedge x)$.

In general, when we wish to check the group action axioms for an *infinite* group, we cannot check axioms GA1 and GA3 from a table – we have to use general algebraic arguments instead.

Exercise 1.7   Consider the group $(\mathbb{Z}, +)$ and the set $X = \{0, 1, 2, 3\}$. Show that

$$n \wedge x = (3n + x) \pmod 4, \quad \text{for } n \in \mathbb{Z} \text{ and } x \in X,$$

defines a group action of $\mathbb{Z}$ on X.

In the next example, the group of real numbers under addition acts on the set of points in the plane. As both the group and the set are infinite, we have to check *all* the axioms by employing general algebraic arguments.

Example 1.5   Show that

$$g \wedge (x, y) = (x + yg, y), \quad \text{for } g \in \mathbb{R} \text{ and } (x, y) \in \mathbb{R}^2,$$

defines a group action of the group $(\mathbb{R}, +)$ on the plane $\mathbb{R}^2$.

Solution   We show that axioms GA1, GA2 and GA3 hold.

GA1 CLOSURE   For each $g \in \mathbb{R}$ and each $(x, y) \in \mathbb{R}^2$, the element $(x + yg, y) \in \mathbb{R}^2$, so axiom GA1 holds.

GA2 IDENTITY   For each $(x, y) \in \mathbb{R}^2$,

$$0 \wedge (x, y) = (x + 0y, y) = (x, y),$$

so axiom GA2 holds.

GA3 COMPOSITION   Let g_1 and g_2 be any elements of $\mathbb{R}$ and let (x, y) be any element of $\mathbb{R}^2$. We must show that

$$g_1 \wedge (g_2 \wedge (x, y)) = (g_1 + g_2) \wedge (x, y).$$

Now

$$\begin{aligned}
g_1 \wedge (g_2 \wedge (x, y)) &= g_1 \wedge (x + yg_2, y) \\
&= (x + yg_2 + yg_1, y) \quad\quad (1.3)
\end{aligned}$$

and

$$\begin{aligned}
(g_1 + g_2) \wedge (x, y) &= (x + y(g_1 + g_2), y) \\
&= (x + yg_2 + yg_1, y). \quad\quad (1.4)
\end{aligned}$$

Expressions (1.3) and (1.4) are the same, so axiom GA3 holds.

Hence $\wedge$ satisfies axioms GA1, GA2 and GA3, so it is a group action. ■

Our final example involves the action of a finite group on an infinite set. We have already seen three natural group actions of $S(\square)$: its action on the vertices of the square, its action on the lines of symmetry of the square (the parts of the lines of symmetry that lie within the square) and its action on the edges of the square. In each of these cases, the elements of the set on which $S(\square)$ acts are subsets of the square. In the next example we show that the natural action of $S(\square)$ on the set of *all* subsets of the square is also a group action.

See the discussion after Exercise 1.4 and see Exercise 1.5(b).

Example 1.6   Let X be the set of all subsets of the square. Show that

$$g \wedge A = g(A), \quad \text{for any subset } A \in X,$$

defines a group action of $S(\square)$ on X.

An illustration of the action of elements of $S(\square)$ on a particular element of X is given below.

If $A =$  , then $a \wedge A =$ and $r \wedge A =$.

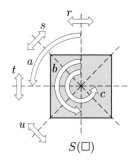

$S(\square)$

Solution   Each element of $S(\square)$ is a symmetry of the square and so maps any subset of the square to a subset of the square; it is thus a function from X to itself. As the group operation is composition of these functions, it follows from Theorem 1.1 that $\wedge$ is a group action of $S(\square)$ on X. ■

Exercise 1.8   With G and X as in Example 1.6, determine the subset $g \wedge A$ for each $g \in S(\square)$, when A is as follows.

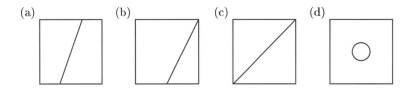

1.4 Actions of matrix groups

We finish this section by looking at some more infinite group actions. The examples in this subsection involve actions of groups of 2×2 matrices on the plane $\mathbb{R}^2$.

We denote the group of all invertible 2×2 matrices under matrix multiplication by M. The matrices in M have a natural action on the points in the plane $\mathbb{R}^2$, given by matrix multiplication. For example, under this action, the matrix

$$\begin{pmatrix} a & b \\ c & d \end{pmatrix}$$

sends the point (x, y) of the plane to the point $(ax + by, cx + dy)$, because

$$\begin{pmatrix} a & b \\ c & d \end{pmatrix} \begin{pmatrix} x \\ y \end{pmatrix} = \begin{pmatrix} ax + by \\ cx + dy \end{pmatrix}.$$

This action is a group action, as we state formally in the following theorem.

Theorem 1.2   Let M be the group of all invertible 2×2 matrices under matrix multiplication. Then

$$\begin{pmatrix} a & b \\ c & d \end{pmatrix} \wedge (x, y) = (ax + by, cx + dy)$$

defines a group action of M on $\mathbb{R}^2$.

By writing the elements of $\mathbb{R}^2$ as ordered pairs, rather than column vectors, and by using $\wedge$, we have disguised the fact that the defined action is matrix multiplication; but that is what it is.

It is straightforward to prove Theorem 1.2 using standard properties of matrix multiplication: it follows immediately from the definition of $\wedge$ that axiom GA1 is satisfied, and that axioms GA2 and GA3 follow, respectively, from the properties of the identity matrix and the associative property of matrix multiplication. We omit the details.

Alternatively, since multiplying an element of $\mathbb{R}^2$ by a 2×2 matrix is the same as applying a linear transformation, the action of M on $\mathbb{R}^2$ is essentially that of a group of functions, so Theorem 1.2 follows from Theorem 1.1. Again we omit the details.

You studied the relationships between matrices and linear transformations in the Linear Algebra Block.

Matrix multiplication is just one of many ways in which a group of 2×2 matrices can act on the plane $\mathbb{R}^2$. We now look at some other examples.

Example 1.7   Consider the multiplicative group of matrices

$$G = \left\{ \begin{pmatrix} a & b \\ 0 & a \end{pmatrix} : a, b \in \mathbb{R}, \ a \neq 0 \right\}.$$

Show that

$$\begin{pmatrix} a & b \\ 0 & a \end{pmatrix} \wedge (x, y) = (ax, ay)$$

defines a group action of G on $\mathbb{R}^2$.

This set was shown to be a group under matrix multiplication in Unit GTB1, Exercise 5.2(b) (where it was denoted by N).

Solution   We show that axioms GA1, GA2 and GA3 hold.

GA1 CLOSURE   Since (ax, ay) is always an element of $\mathbb{R}^2$, axiom GA1 holds.

GA2 IDENTITY   The identity element of G is $\begin{pmatrix} 1 & 0 \\ 0 & 1 \end{pmatrix}$, and

$$\begin{pmatrix} 1 & 0 \\ 0 & 1 \end{pmatrix} \wedge (x, y) = (1x, 1y) = (x, y),$$

for all $(x, y) \in \mathbb{R}^2$, so axiom GA2 holds.

GA3 COMPOSITION   Let $\begin{pmatrix} a & b \\ 0 & a \end{pmatrix}, \begin{pmatrix} c & d \\ 0 & c \end{pmatrix} \in G$ and $(x, y) \in \mathbb{R}^2$. We must show that

$$\begin{pmatrix} a & b \\ 0 & a \end{pmatrix} \wedge \left(\begin{pmatrix} c & d \\ 0 & c \end{pmatrix} \wedge (x, y) \right)$$

$$= \left(\begin{pmatrix} a & b \\ 0 & a \end{pmatrix} \begin{pmatrix} c & d \\ 0 & c \end{pmatrix} \right) \wedge (x, y).$$

We have

$$\begin{pmatrix} a & b \\ 0 & a \end{pmatrix} \wedge \left(\begin{pmatrix} c & d \\ 0 & c \end{pmatrix} \wedge (x, y) \right)$$

$$= \begin{pmatrix} a & b \\ 0 & a \end{pmatrix} \wedge (cx, cy)$$

$$= (acx, acy) \tag{1.5}$$

and

$$\left(\begin{pmatrix} a & b \\ 0 & a \end{pmatrix} \begin{pmatrix} c & d \\ 0 & c \end{pmatrix} \right) \wedge (x, y)$$

$$= \begin{pmatrix} ac & ad + bc \\ 0 & ac \end{pmatrix} \wedge (x, y)$$

$$= (acx, acy). \tag{1.6}$$

Expressions (1.5) and (1.6) are the same, so axiom GA3 holds.

Hence $\wedge$ satisfies axioms GA1, GA2 and GA3, so it is a group action.   ∎

Exercise 1.9 Consider the multiplicative group of matrices

$$G = \left\{ \begin{pmatrix} a & b \\ 0 & 1 \end{pmatrix} : a, b \in \mathbb{R}, \ a \neq 0 \right\}.$$

You may assume that this set is a group under matrix multiplication.

Which of the following defines a group action of G on $\mathbb{R}^2$? In each case, justify your answer by giving either a proof or a counter-example.

(a) $\begin{pmatrix} a & b \\ 0 & 1 \end{pmatrix} \wedge (x, y) = (ax, y)$

(b) $\begin{pmatrix} a & b \\ 0 & 1 \end{pmatrix} \wedge (x, y) = (ax, by)$

(c) $\begin{pmatrix} a & b \\ 0 & 1 \end{pmatrix} \wedge (x, y) = (ax, by + y)$

Further exercises

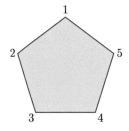

Exercise 1.10 Representing each of the 10 elements of $S(\bigcirc)$ as a permutation of the vertex labels 1, 2, 3, 4, 5, construct the action table for the natural action of $S(\bigcirc)$ on $\{1, 2, 3, 4, 5\}$.

Exercise 1.11 For each of the following cases, construct an action table for $\wedge$ and determine whether $\wedge$ defines a group action of G on X.

(a) $G = \mathbb{Z}_6$ and $X = \{0, 1, 2\}$, with $\wedge$ defined by

$$g \wedge x = g +_3 x, \quad \text{for } g \in \mathbb{Z}_6 \text{ and } x \in \{0, 1, 2\}.$$

Remember that the binary operation in $\mathbb{Z}_6$ is $+_6$, and that $g +_3 x$ is the remainder when $g + x$ is divided by 3.

(b) $G = \mathbb{Z}_6$ and $X = \{0, 1, 2, 3\}$, with $\wedge$ defined by

$$g \wedge x = g +_4 x, \quad \text{for } g \in \mathbb{Z}_6 \text{ and } x \in \{0, 1, 2, 3\}.$$

Exercise 1.12 Let X be the set of all unordered pairs of symbols from $\{1, 2, 3, 4\}$; that is,

$$X = \{\{1, 2\}, \{1, 3\}, \{1, 4\}, \{2, 3\}, \{2, 4\}, \{3, 4\}\}.$$

The group S_4 acts on X in a natural way; for example,

$$(1\ 3\ 4) \wedge \{1, 3\} = \{3, 4\} \quad \text{(since } (1\ 3\ 4) \text{ maps 1 to 3 and 3 to 4)}$$

and

$$(2\ 4) \wedge \{2, 3\} = \{3, 4\} \quad \text{(since } (2\ 4) \text{ maps 2 to 4 and fixes 3).}$$

By checking axioms GA1, GA2 and GA3, show that $\wedge$ is a group action of S_4 on X.

Remember that $\{3, 4\} = \{4, 3\}$; there is no ordering on the elements in each pair.

Exercise 1.13 Let $\wedge$ be defined by

$$r \wedge (x, y) = (x, y + r), \quad \text{for } r \in \mathbb{R} \text{ and } (x, y) \in \mathbb{R}^2.$$

Show that $\wedge$ is a group action of $(\mathbb{R}, +)$ on $\mathbb{R}^2$.

2   Orbits and stabilisers

After working through this section, you should be able to:

(a) explain what is meant by the *orbit* of an element of X and the *stabiliser* of an element of X, when a group G acts on a set X;

(b) understand that the orbits of a group action form a partition of the set X;

(c) understand that the stabiliser of an element of X is a subgroup of the group G;

(d) determine orbits and stabilisers for a given group action.

2.1 Orbits and stabilisers

In this audio section we introduce the concepts of *orbits* and *stabilisers* for a group action. Throughout the section, X is a set and G is a group acting on X. For each element $x \in X$, we define the orbit of x, which is a subset of X, and we see how the orbits form a partition of the set X. We also define the stabiliser of x, which is a subgroup of G. Later in the unit we show that the orbit and stabiliser of x are, in a sense, complementary notions.

Listen to the audio as you work through the frames.

Audio

1. Introducing orbits

Example 2.1

$G = S(\square)$

$X = \{1, 2, 3, 4\}$

```
   4        1
   
   
   3        2
```

1 can be sent to 1, 2, 3 or 4.

We say

$\qquad Orb(1) = \{1, 2, 3, 4\}$;

similarly,

$\qquad Orb(2) = \{1, 2, 3, 4\}$

$\qquad = Orb(3) = Orb(4) = X.$

^	**1**	2	3	4
e	1	2	3	4
a	**2**	3	4	1
b	3	4	1	2
c	4	1	2	3
r	4	3	2	1
s	1	4	3	2
t	2	1	4	3
u	3	2	1	4

Example 2.2

$G = S(\square)$

$X = \{R, S, T, U\}$

R can be sent to R and T:

$\qquad Orb(R) = \{R, T\} = Orb(T)$;

similarly,

$\qquad Orb(S) = \{S, U\} = Orb(U)$.

^	R	S	T	U
e	R	S	T	U
a	T	U	R	S
b	R	S	T	U
c	T	U	R	S
r	R	U	T	S
s	T	S	R	U
t	R	U	T	S
u	T	S	R	U

2. Orbits

Definition   Let a group G act on a set X.

For $x \in X$, the **orbit** of x is

$\qquad Orb(x) = \{ g \wedge x : g \in G \}.$

set of all elements of X which can be reached from x

$Orb(x)$ is a subset of X

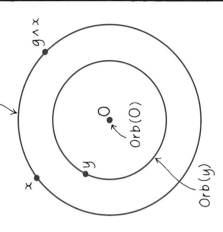

3. An example in $\mathbb{R}^2$

$G = \{\text{rotations about } O\}$

$X = \mathbb{R}^2$

$Orb(O) = \{O\}$.

All other orbits are concentric circles, with centre O.

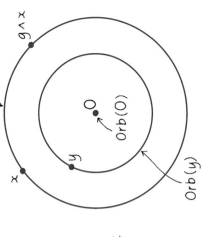

6. $G = \left\{ \begin{pmatrix} a & b \\ 0 & a \end{pmatrix} : a,b \in \mathbb{R}, a \neq 0 \right\}$, $X = \mathbb{R}^2$, $\begin{pmatrix} a & b \\ 0 & a \end{pmatrix} \wedge (x,y) = (ax, ay)$

$\text{Orb}(x,y) = \{(ax, ay) : a \in \mathbb{R}, a \neq 0\}$

$\text{Orb}(0,0) = \{(0,0)\}$

$\text{Orb}(1,2) = \{(a, 2a) : a \in \mathbb{R}, a \neq 0\}$

line $y = 2x$, excluding 0

What is the orbit of (x,y)?

For x, y not both 0,

$\text{Orb}(x,y) = \{(ax, ay) : a \in \mathbb{R}, a \neq 0\}$

line through 0, excluding 0

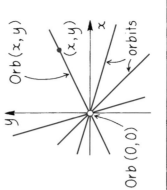

7. Exercise 2.2

Find all the orbits when $G = \left\{ \begin{pmatrix} a & b \\ 0 & 1 \end{pmatrix} : a, b \in \mathbb{R}, a \neq 0 \right\}$,

$X = \mathbb{R}^2$ and $\begin{pmatrix} a & b \\ 0 & 1 \end{pmatrix} \wedge (x,y) = (ax, y)$.

4. $G = \left\{ \begin{pmatrix} a & 0 \\ 0 & b \end{pmatrix} : a,b \in \mathbb{R}^+ \right\}$, $X = \mathbb{R}^2$, $\begin{pmatrix} a & 0 \\ 0 & b \end{pmatrix} \wedge (x,y) = (ax, by)$

$\text{Orb}(x,y) = \{(ax, by) : a, b \in \mathbb{R}^+\}$

so

$\text{Orb}(0,0) = (0,0)$

$\text{Orb}(-1,0) = \{(-a, 0) : a > 0\}$

$\text{Orb}(1,-1) = \{(a, -b) : a, b > 0\}$

Exercise 2.1 Find the remaining 6 orbits.

5. $G = (\mathbb{R}, +)$, $X = \mathbb{R}^2$, $g \wedge (x,y) = (x + yg, y)$

$\text{Orb}(x,y) = \{(x + yg, y) : g \in \mathbb{R}\}$

$\text{Orb}(2,0) = \{(2,0)\}$

$\text{Orb}(x,0) = \{(x,0)\}$

point

$\text{Orb}(1,2) = \{(1 + 2g, 2) : g \in \mathbb{R}\}$

For $y \neq 0$,

$\text{Orb}(x,y) = \{(x + yg, y) : g \in \mathbb{R}\}$

line

8. Partitions of $\mathbb{R}^2$

Frame 3   Frame 4   Frame 5   Frame 6

In each case,

- any two orbits are either identical or disjoint

- the union of the orbits is $X = \mathbb{R}^2$.

9. An important result

Theorem 2.1 Let a group G act on a set X.

Then the orbits partition X.

Outline proof Define a relation $\sim$ on X by:

$x \sim y$ if y is in the orbit of x.

Then

- $\sim$ is an equivalence relation,

- the equivalence classes are the **orbits**, so the orbits form a partition of X.

proof in text

10. Strategy 2.1 For finding orbits

To find all the orbits in X:

1. Choose any $x \in X$, and find Orb(x).

2. Choose any element of X not yet assigned to an orbit, and find its orbit.

3. Repeat step 2 until X is partitioned.

- consider particular elements
- try to spot general pattern

11. Exercise 2.3

Use the above strategy to determine all the orbits for each of the following group actions:

(a) $G = (\mathbb{Z}, +)$, $X = \{0, 1, 2, 3, 4, 5\}$,

$g \wedge x = (2g + x) \pmod 6$;

(b) $G = S(\square)$, $X = \{1, 2, 3, 4\}$,

G acts on X by permuting the vertices.

13. Stabilisers

Definition Let a group G act on a set X.

For $x \in X$, the **stabiliser** of x is

$$\text{Stab}(x) = \{g \in G : g \wedge x = x\}.$$

set of all elements of G which fix x

Stab(x) is a subset of G

14. An example in $\mathbb{R}^2$

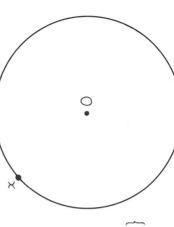

$G = \{\text{rotations about } O\}$

$X = \mathbb{R}^2$

$\text{Stab}(O) = G.$

For any other $x \in X$:

$\text{Stab}(x) = \{0, \pm 2\pi, \pm 4\pi, \ldots\}$

$= \{2n\pi : n \in \mathbb{Z}\}.$

12. Introducing stabilisers

Example 2.3

$G = S(\square)$

$X = \{1, 2, 3, 4\}$

$\wedge$	1	2	3	4
e	**1**	2	3	4
a	2	3	4	1
b	3	4	1	2
c	4	1	2	3
r	4	3	2	1
s	**1**	4	3	2
t	2	1	4	3
u	3	2	1	4

1 is fixed by e and s:

we say $\text{Stab}(1) = \{e, s\}$;

similarly:

$\text{Stab}(2) = \{e, u\},$

$\text{Stab}(3) = \{e, s\},$

$\text{Stab}(4) = \{e, u\}.$

Example 2.4

$G = S(\square)$

$X = \{R, S, T, U\}$

$\wedge$	R	S	T	U
e	**R**	S	T	U
a	T	U	R	S
b	**R**	S	T	U
c	T	U	R	S
r	**R**	U	T	S
s	T	S	R	U
t	**R**	U	T	S
u	T	S	R	U

R is fixed by e, b, r and t,

so $\text{Stab}(R) = \{e, b, r, t\}$;

similarly:

$\text{Stab}(T) = \{e, b, r, t\},$

$\text{Stab}(S) = \{e, b, s, u\},$

$\text{Stab}(U) = \{e, b, s, u\}.$

17. $G = \left\{ \begin{pmatrix} a & b \\ 0 & a \end{pmatrix} : a, b \in \mathbb{R}, a \neq 0 \right\}$, $X = \mathbb{R}^2$, $\begin{pmatrix} a & b \\ 0 & a \end{pmatrix} \wedge (x, y) = (ax, ay)$

$\text{Stab}(x, y) = \left\{ \begin{pmatrix} a & b \\ 0 & a \end{pmatrix} : \begin{pmatrix} a & b \\ 0 & a \end{pmatrix} \wedge (x, y) = (x, y), a, b \in \mathbb{R}, a \neq 0 \right\}$

$= \left\{ \begin{pmatrix} a & b \\ 0 & a \end{pmatrix} : (ax, ay) = (x, y), a, b \in \mathbb{R}, a \neq 0 \right\}.$

(cloud: $b \in \mathbb{R}$)

- If $x = y = 0$, then a can be any non-zero real number,

  so $\text{Stab}(0, 0) = G$.

- If $x \neq 0$ or $y \neq 0$, a must be 1,

  so $\text{Stab}(x, y) = \left\{ \begin{pmatrix} 1 & b \\ 0 & 1 \end{pmatrix} : b \in \mathbb{R} \right\}.$

18. An important result

Theorem 2.2   Let a group G act on a set X.

Then, for each $x \in X$,

Stab (x) is a subgroup of G.

(cloud: proof in text)

19. Exercise 2.4

For each of the group actions in Exercises 2.1 – 2.3 determine the stabiliser of each element of X.

(cloud: Frames 4, 7, 11)

15. $G = \left\{ \begin{pmatrix} a & 0 \\ 0 & b \end{pmatrix} : a, b \in \mathbb{R}^+ \right\}$, $X = \mathbb{R}^2$, $\begin{pmatrix} a & 0 \\ 0 & b \end{pmatrix} \wedge (x, y) = (ax, by)$

$\text{Stab}(x, y) = \left\{ \begin{pmatrix} a & 0 \\ 0 & b \end{pmatrix} : \begin{pmatrix} a & 0 \\ 0 & b \end{pmatrix} \wedge (x, y) = (x, y), a, b \in \mathbb{R}^+ \right\}$

$= \left\{ \begin{pmatrix} a & 0 \\ 0 & b \end{pmatrix} : (ax, by) = (x, y) \right\}.$

so

$\text{Stab}(0, 0) = \left\{ \begin{pmatrix} a & 0 \\ 0 & b \end{pmatrix} : (a0, b0) = (0, 0) \right\} = G;$

$\text{Stab}(-1, 0) = \left\{ \begin{pmatrix} a & 0 \\ 0 & b \end{pmatrix} : (-a, 0) = (-1, 0) \right\} = \left\{ \begin{pmatrix} 1 & 0 \\ 0 & b \end{pmatrix} : b \in \mathbb{R}^+ \right\};$

$\text{Stab}(1, -1) = \left\{ \begin{pmatrix} a & 0 \\ 0 & b \end{pmatrix} : (a, -b) = (1, -1) \right\} = \left\{ \begin{pmatrix} 1 & 0 \\ 0 & 1 \end{pmatrix} \right\}.$

16. $G = (\mathbb{R}, +)$, $X = \mathbb{R}^2$, $g \wedge (x, y) = (x + yg, y)$

What is the stabiliser of a general point (x, y)?

$\text{Stab}(x, y) = \{ g : g \wedge (x, y) = (x, y) \};$

$g \wedge (x, y) = (x, y)$, so $(x + yg, y) = (x, y).$

Thus $x + yg = x$, so $yg = 0$.

(cloud: $y = 0$ or $g = 0$)

- If $y = 0$, g can be any real number.

- If $y \neq 0$, g must be 0.

So $\text{Stab}(x, y) = \begin{cases} \mathbb{R}, & \text{if } y = 0, \\ \{0\}, & \text{if } y \neq 0. \end{cases}$

23

Post-audio exercise

We conclude this section with an additional exercise on orbits and stabilisers. The solution of this exercise will be needed at the beginning of Section 3.

Exercise 2.5   Let $G = S(\square)$ and let X be the set of all subsets of the square, as in Example 1.6. For each of the elements A of X in the following table, find $\mathrm{Orb}(A)$ and $\mathrm{Stab}(A)$, and complete the table.

$|\mathrm{Orb}(A)|$ denotes the number of elements in $\mathrm{Orb}(A)$.

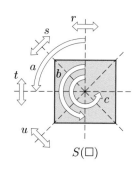

$S(\square)$

Subset A	$\mathrm{Orb}(A)$	$\mathrm{Stab}(A)$	$|\mathrm{Orb}(A)|$	$|\mathrm{Stab}(A)|$
▧	$\{▧, ▨, ◣, ◪\}$	$\{e, b\}$	4	2
▨	?	?	?	?
◸	?	?	?	?
◎	?	?	?	?

Comment on the relative sizes of $\mathrm{Orb}(A)$ and $\mathrm{Stab}(A)$.

2.2 Proofs

In the audio section we stated two theorems without proof. We now supply these proofs.

Theorem 2.1   Let a group G act on a set X. Then the orbits form a partition of X.

Proof   We define a relation $\sim$ on X as follows:

$$x \sim y \quad \text{if} \quad y = g \wedge x, \text{ for some } g \in G.$$

This says that $x \sim y$ if $y \in \mathrm{Orb}(x)$.

First we prove that $\sim$ is an equivalence relation. We show that $\sim$ has the reflexive, symmetric and transitive properties.

E1 REFLEXIVE   Let $x \in X$. Then $x = e \wedge x$, by axiom GA2 for a group action. Thus $x \sim x$, so the relation is reflexive.

Is $x \in \mathrm{Orb}(x)$?

E2 SYMMETRIC   Let x and y be any elements of X with $x \sim y$. Then

$$y = g \wedge x, \quad \text{for some } g \in G.$$

Letting the element $g^{-1} \in G$ act on both sides of this equation gives

$$\begin{aligned}
g^{-1} \wedge y &= g^{-1} \wedge (g \wedge x) \\
&= (g^{-1} \circ g) \wedge x \quad \text{(axiom GA3)} \\
&= e \wedge x \\
&= x \quad \text{(axiom GA2)}.
\end{aligned}$$

If $y \in \mathrm{Orb}(x)$, is $x \in \mathrm{Orb}(y)$?

This shows that

if $y = g \wedge x$, then $x = g^{-1} \wedge y$
This is a useful result that is worth remembering.

Since $x = g^{-1} \wedge y$ and $g^{-1} \in G$, we have $y \sim x$, so the relation is symmetric.

If $y \in \mathrm{Orb}(x)$ and $z \in \mathrm{Orb}(y)$, is $z \in \mathrm{Orb}(x)$?

E3 TRANSITIVE   Let x, y and z be any elements of X such that $x \sim y$ and $y \sim z$. Then

$$y = g \wedge x, \quad \text{for some element } g \in G$$

and

$$z = h \wedge y, \quad \text{for some element } h \in G.$$

It follows that

$$\begin{aligned} z &= h \wedge y \\ &= h \wedge (g \wedge x) \\ &= (h \circ g) \wedge x \quad \text{(axiom GA3)}. \end{aligned}$$

As $h \circ g \in G$, we deduce that $x \sim z$, so the relation is transitive.

Hence $\sim$ has the reflexive, symmetric and transitive properties, so it is an equivalence relation on X.

The equivalence classes of any equivalence relation form a partition of the set on which the relation is defined. Thus the equivalence classes of $\sim$ form a partition of X. But, from the definition of $\sim$, the equivalence class of $x \in X$ is

$$\{y \in X : y = g \wedge x, \text{ for some } g \in G\} = \text{Orb}(x).$$

Thus the equivalence classes are precisely the orbits, so the orbits form a partition of X. ∎

Theorem 2.2 Let a group G act on a set X. Then, for each $x \in X$, the set $\text{Stab}(x)$ is a subgroup of G.

Proof   We show that the three subgroup properties hold.

SG1 CLOSURE   Let $g, h \in \text{Stab}(x)$; then

$$g \wedge x = x \quad \text{and} \quad h \wedge x = x.$$

It follows that

$$\begin{aligned} (g \circ h) \wedge x &= g \wedge (h \wedge x) \quad \text{(axiom GA3)} \\ &= g \wedge x \\ &= x. \end{aligned}$$

Thus $g \circ h \in \text{Stab}(x)$, so property SG1 holds.

SG2 IDENTITY   Since $e \wedge x = x$, by axiom GA2, it follows that $e \in \text{Stab}(x)$, so property SG2 holds.

SG3 INVERSES   Let $g \in \text{Stab}(x)$; then

$$g \wedge x = x.$$

It follows that

$$\begin{aligned} g^{-1} \wedge x &= g^{-1} \wedge (g \wedge x) \\ &= (g^{-1} \circ g) \wedge x \quad \text{(axiom GA3)} \\ &= e \wedge x \\ &= x \quad \text{(axiom GA2)}. \end{aligned}$$

Thus $g^{-1} \in \text{Stab}(x)$, so property SG3 holds.

Since properties SG1, SG2 and SG3 hold, $\text{Stab}(x)$ is a subgroup of G. ∎

The elements of $\text{Stab}(x)$ are said to *stabilise* x.

Further exercises

Exercise 2.6   The following table is an action table for a group G of order 8 acting on a set X of 7 elements.

$\wedge$	1	2	3	4	5	6	7
e	1	2	3	4	5	6	7
a	3	4	5	2	7	6	1
b	5	2	7	4	1	6	3
c	7	4	1	2	3	6	5
w	3	4	1	2	7	6	5
x	5	2	3	4	1	6	7
y	7	4	5	2	3	6	1
z	1	2	7	4	5	6	3

(a) Write down the partition of X into orbits.

(b) Write down the stabiliser of each of the elements of X.

(c) For each $x \in X$, verify that

$$|\mathrm{Orb}(x)| \times |\mathrm{Stab}(x)| = |G|.$$

Exercise 2.7   For the group action in Exercise 1.12 (page 17), show that there is only one orbit, and determine $\mathrm{Stab}\,(\{i, j\})$ for each $\{i, j\} \in X$.

Exercise 2.8   For the group action in Exercise 1.13, show that, for each $(x, y) \in \mathbb{R}^2$, the set $\mathrm{Stab}\,(x, y)$ consists of the identity element alone. Determine the set of orbits of this action.

Exercise 2.9   Let the 10 elements of $S(\bigcirc)$ be represented as permutations of the five vertex labels 1, 2, 3, 4, 5, as in Exercise 1.10. Let X be the set of 10 unordered pairs of elements from $\{1, 2, 3, 4, 5\}$; that is,

$$X = \{\{1, 2\}, \{1, 3\}, \{1, 4\}, \{1, 5\}, \{2, 3\}, \{2, 4\}, \{2, 5\},$$
$$\{3, 4\}, \{3, 5\}, \{4, 5\}\}.$$

Let $S(\bigcirc)$ act on X in the natural way; for example,

$$(1\ 2\ 3\ 4\ 5) \wedge \{2, 4\} = \{3, 5\},$$

since $(1\ 2\ 3\ 4\ 5)$ maps 2 to 3 and 4 to 5.

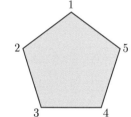

(a) Determine the partition of X into orbits of this action.

(b) Determine $\mathrm{Stab}\,(\{1, 2\})$ and $\mathrm{Stab}\,(\{1, 3\})$.

3   The Orbit–Stabiliser Theorem

After working through this section, you should be able to:

(a) understand the *Orbit–Stabiliser Theorem*;

(b) determine the correspondence between the left cosets of $\mathrm{Stab}(x)$ and the elements of $\mathrm{Orb}(x)$, where x is an element of a set being acted on;

(c) understand various ways in which a group can act on itself.

3.1 Statement of the Orbit–Stabiliser Theorem

In Exercise 2.5 we considered the natural group action of $S(\square)$ on the set X of all subsets of the square. We found the orbits and stabilisers of some particular subsets of the square under this action. We obtained the following table.

Subset A	Orb(A)	Stab(A)	|Orb(A)|	|Stab(A)|
▱	{▱,◱,◲,◳}	$\{e, b\}$	4	2
◩	{◩,◪,◨,◧, ◫,◩,◨,◧}	$\{e\}$	8	1
◪	{◪,◩}	$\{e, b, s, u\}$	2	4
⊙	{⊙}	$S(\square)$	1	8

From the last two columns of this table, you can see that, in each case,

$$|\mathrm{Orb}(A)| \times |\mathrm{Stab}(A)| = 8.$$

That is, for each of the chosen subsets of the square, the number of elements in its orbit, multiplied by the number of elements in its stabiliser, is equal to the order of the group. These facts are instances of the following general result, which is the main result of this section.

Theorem 3.1 Orbit–Stabiliser Theorem

Let a finite group G act on a set X. Then, for each $x \in X$,

$$|\mathrm{Orb}(x)| \times |\mathrm{Stab}(x)| = |G|.$$

We prove this theorem later in the section.

If a finite group G acts on a set X and $x \in X$, then, since $\mathrm{Stab}(x)$ is a subgroup of G, we know from Lagrange's Theorem that the order of $\mathrm{Stab}(x)$ divides the order of G. The Orbit–Stabiliser Theorem confirms this, and tells us more. Rearranging the equation in the Orbit–Stabiliser Theorem gives

$$\frac{|G|}{|\mathrm{Stab}(x)|} = |\mathrm{Orb}(x)|.$$

This equation tells us that the index of $\mathrm{Stab}(x)$ in G—that is, the number of left cosets of the subgroup $\mathrm{Stab}(x)$ in G—is equal to the number of elements in the orbit of x.

Recall that, if H is a subgroup of a finite group G, then the index of H in G is equal to $|G|/|H|$.

The following corollary of the Orbit–Stabiliser Theorem is of particular interest.

Corollary to Theorem 3.1

Let a finite group G act on a set X. Then, for each $x \in X$, the number of elements in $\mathrm{Orb}(x)$ divides the order of G.

For example, in the table at the beginning of this subsection, the orbits have 1, 2, 4 or 8 elements, and these numbers all divide 8, the order of $S(\square)$.

Exercise 3.1   Verify the statement of the Orbit–Stabiliser Theorem, and its corollary, for each $x \in X$ in each of the following group actions:

(a) $S(\square)$ acting on $X = \{1, 2, 3, 4\}$, the set of vertices of the square;

The group action is the natural one in each case.

(b) $S(\square)$ acting on $X = \{R, S, T, U\}$, the four lines of symmetry of the square;

(c) $S(\triangle)$ acting on $X = \{R, S, T\}$, the three lines of symmetry of the triangle.

3.2 Stabilisers and cosets

For any group action, the left cosets of any stabiliser have an important property, which we meet in this subsection. Here is an example.

Consider the natural action of the group S_3 on the set of symbols $\{1, 2, 3\}$. The action table is reproduced in the margin. Let us find the stabiliser of the element 1. By looking down the column labelled 1, we see that two elements send 1 to itself, namely e and $(2\ 3)$. So

$$\text{Stab}(1) = \{e, (2\ 3)\}.$$

The left cosets of $\text{Stab}(1)$ are

$$\text{Stab}(1) = \{e, (2\ 3)\},$$
$$(1\ 2)\,\text{Stab}(1) = \{(1\ 2) \circ e, (1\ 2) \circ (2\ 3)\} = \{(1\ 2), (1\ 2\ 3)\},$$
$$(1\ 3)\,\text{Stab}(1) = \{(1\ 3) \circ e, (1\ 3) \circ (2\ 3)\} = \{(1\ 3), (1\ 3\ 2)\}.$$

$\wedge$	1	2	3
e	1	2	3
$(1\ 2\ 3)$	2	3	1
$(1\ 3\ 2)$	3	1	2
$(1\ 2)$	2	1	3
$(1\ 3)$	3	2	1
$(2\ 3)$	1	3	2

By definition, each element of the first coset, $\text{Stab}(1)$ itself, sends 1 to 1. Let us consider the action on 1 of the elements in the other two cosets. For the second coset, $(1\ 2)\,\text{Stab}(1)$, we have

$$(1\ 2) \wedge 1 = 2, \quad (1\ 2\ 3) \wedge 1 = 2,$$

so each element of this coset sends 1 to 2. For the third coset, $(1\ 3)\,\text{Stab}(1)$, we have

$$(1\ 3) \wedge 1 = 3, \quad (1\ 3\ 2) \wedge 1 = 3,$$

so each element of this coset sends 1 to 3.

So if two group elements lie in the *same* left coset of $\text{Stab}(1)$, then they send 1 to the *same* element of X; whereas if they lie in *different* left cosets, then they send 1 to *different* elements of X.

In the next exercise you are asked to check whether the left cosets of Stab(2) in the same group action satisfy a similar property.

Exercise 3.2   Use the action table in the margin at the beginning of this subsection to do the following.

(a) Find Stab(2).

(b) Find the left cosets of Stab(2).

(c) Taking each left coset in turn, write down $g \wedge 2$ for each element g of the coset.

Check that group elements in the same left coset of Stab(2) send 2 to the same element of X, and that group elements in different left cosets send 2 to different elements of X.

The examples above illustrate the following general result.

Theorem 3.2   Let $\wedge$ be a group action of G on X, let $x \in X$, and let g and h be any elements of G. Then

$$g \wedge x = h \wedge x$$

if and only if

g and h lie in the same left coset of Stab(x).

This theorem involves left, rather than right, cosets because of the way we defined a group action. Stab(x) may not be a normal subgroup of G, so left and right cosets may be different.

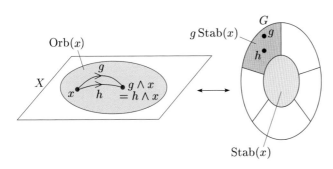

Proof   First we prove the 'if' part. Suppose that

g and h lie in the same left coset of Stab(x).

Then $h \in g\,\mathrm{Stab}(x)$, so $h = g \circ k$, where $k \in \mathrm{Stab}(x)$. It follows that

$$\begin{aligned}
h \wedge x &= (g \circ k) \wedge x \\
&= g \wedge (k \wedge x) \quad \text{(axiom GA3)} \\
&= g \wedge x \quad (k \in \mathrm{Stab}(x)),
\end{aligned}$$

as required.

Now we prove the 'only if' part. Suppose that

$$g \wedge x = h \wedge x.$$

Consider the action of the group element $g^{-1} \circ h$ on x:

$$\begin{aligned}
(g^{-1} \circ h) \wedge x &= g^{-1} \wedge (h \wedge x) \quad \text{(axiom GA3)} \\
&= g^{-1} \wedge (g \wedge x) \\
&= (g^{-1} \circ g) \wedge x \quad \text{(axiom GA3)} \\
&= e \wedge x \\
&= x \quad \text{(axiom GA2).}
\end{aligned}$$

Therefore $g^{-1} \circ h = k$, for some $k \in \mathrm{Stab}(x)$. It follows, by composing on the left by g, that $h = g \circ k$. Hence $h \in g\,\mathrm{Stab}(x)$. Thus g and h lie in the same left coset of Stab(x). ∎

If you are short of time, omit this proof.

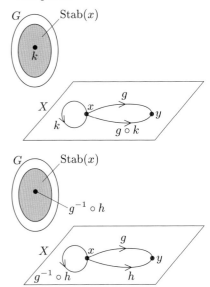

Theorem 3.2 shows that if G is a group acting on a set X and x is any element of X, then the sets of group elements that send x to a common element of X are precisely the left cosets of $\mathrm{Stab}(x)$. This means that

if we collect together the group elements according to where they send x, then we have the left cosets of $\mathrm{Stab}(x)$;

and that, conversely,

if we find the left cosets of $\mathrm{Stab}(x)$, then we have the sets of group elements that send x to the same element of X.

Exercise 3.3   Consider the group action of $S(\square)$ on the set $\{1,2,3,4\}$ of vertices of the square.

(a) Find $\mathrm{Stab}(4)$ and the left cosets of $\mathrm{Stab}(4)$ in $S(\square)$.

(b) Find the partition of $S(\square)$ obtained by collecting together elements of $S(\square)$ that send 4 to the same element of $\{1,2,3,4\}$. Check that this partition is the same as that in part (a).

Theorem 3.2 has the following corollary.

$\wedge$	1	2	3	4
e	1	2	3	4
a	2	3	4	1
b	3	4	1	2
c	4	1	2	3
r	4	3	2	1
s	1	4	3	2
t	2	1	4	3
u	3	2	1	4

$S(\square)$ acting on $\{1,2,3,4\}$

Corollary to Theorem 3.2

Let $\wedge$ be a group action of G on X and let $x \in X$. Then there is a one-one correspondence between the left cosets of $\mathrm{Stab}(x)$ in G and the elements of $\mathrm{Orb}(x)$, given by

$$g\,\mathrm{Stab}(x) \longleftrightarrow g \wedge x.$$

$\circ$	e	a	b	c	r	s	t	u
e	e	a	b	c	r	s	t	u
a	a	b	c	e	s	t	u	r
b	b	c	e	a	t	u	r	s
c	c	e	a	b	u	r	s	t
r	r	u	t	s	e	c	b	a
s	s	r	u	t	a	e	c	b
t	t	s	r	u	b	a	e	c
u	u	t	s	r	c	b	a	e

$S(\square)$

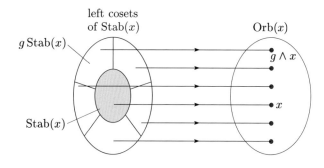

Proof   We define a function f as follows:

$$f : \{\text{left cosets of } \mathrm{Stab}(x)\} \longrightarrow \mathrm{Orb}(x)$$
$$g\,\mathrm{Stab}(x) \longmapsto g \wedge x.$$

That is, f maps each left coset of $\mathrm{Stab}(x)$ to $g \wedge x$, where g is any element of the coset. This is a valid definition of a function because, by Theorem 3.2, all elements of a left coset of $\mathrm{Stab}(x)$ send x to the *same* element of X.

Theorem 3.2 also tells us that elements from *different* left cosets of $\mathrm{Stab}(x)$ send x to *different* elements of X, so f is one-one.

Also, f is onto because each element $g \wedge x$ of $\mathrm{Orb}(x)$ is the image under f of the coset $g\,\mathrm{Stab}(x)$.

Hence f is a one-one correspondence.   ∎

To illustrate the correspondence given by the corollary to Theorem 3.2, consider again the natural action of S_3 on the set $\{1,2,3\}$ that we considered at the beginning of this subsection.

We looked at the element 1 of $\{1, 2, 3\}$. In this case the correspondence is as follows.

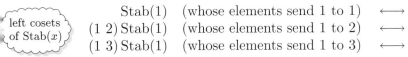

$$
\begin{array}{ccccc}
\text{left cosets} & \text{Stab}(1) & \text{(whose elements send 1 to 1)} & \longleftrightarrow & 1 \\
\text{of Stab}(x) & (1\ 2)\,\text{Stab}(1) & \text{(whose elements send 1 to 2)} & \longleftrightarrow & 2 \quad \text{elements of} \\
 & (1\ 3)\,\text{Stab}(1) & \text{(whose elements send 1 to 3)} & \longleftrightarrow & 3 \quad \text{Orb}(1)
\end{array}
$$

As a further illustration of the correspondence in the corollary, consider again the group action in Example 1.4. This is an action of the group $(\mathbb{Z}, +)$ on the set $X = \{0, 1, 2, 3, 4, 5\}$, given by

$$n \wedge x = (2n + x) \pmod 6, \quad \text{for } n \in \mathbb{Z} \text{ and } x \in X.$$

Let us look at the element 2 of X. The action of any group element n on 2 is given by

$$n \wedge 2 = (2n + 2) \pmod 6.$$

Part of the column of the action table corresponding to $x = 2$ is shown in the margin. We see that $\text{Orb}(2) = \{0, 2, 4\}$ and that $\text{Stab}(2)$ consists of all the multiples of 3. That is,

$$\text{Stab}(2) = \{\ldots, -6, -3, 0, 3, 6, \ldots\} = 3\mathbb{Z}.$$

The cosets of $\text{Stab}(2)$ are as follows. (The group $(\mathbb{Z}, +)$ is Abelian, so left and right cosets coincide in this case.)

$$
\begin{aligned}
3\mathbb{Z} &= \{\ldots, -6, -3, 0, 3, 6, \ldots\}, \\
1 + 3\mathbb{Z} &= \{\ldots, -5, -2, 1, 4, 7, \ldots\}, \\
2 + 3\mathbb{Z} &= \{\ldots, -4, -1, 2, 5, 8, \ldots\}.
\end{aligned}
$$

$\wedge$	$\cdots$ 2 $\cdots$
$\vdots$	$\vdots$
-5	4
-4	0
-3	2
-2	4
-1	0
0	2
1	4
2	0
3	2
4	4
5	0
$\vdots$	$\vdots$

We can see from the partial action table in the margin that, as expected from Theorem 3.2, all the elements of the coset $1 + 3\mathbb{Z}$ send 2 to the same element of X, namely 4, and all the elements of the coset $2 + 3\mathbb{Z}$ send 2 to the same element of X, namely 0. Thus the correspondence of the corollary to Theorem 3.2 is as follows.

$$
\begin{array}{ccc}
 & 3\mathbb{Z} & \longleftrightarrow \quad 2 \\
\text{cosets of} & 1 + 3\mathbb{Z} & \longleftrightarrow \quad 4 \quad \text{elements of} \\
\text{Stab}(2) = 3\mathbb{Z} & 2 + 3\mathbb{Z} & \longleftrightarrow \quad 0 \quad \text{Orb}(2)
\end{array}
$$

Exercise 3.4   Write down the one-one correspondence between the left cosets of $\text{Stab}(4)$ and the elements of $\text{Orb}(4)$ in the group action of $S(\square)$ on the set $\{1, 2, 3, 4\}$ of vertices of the square.

Hint: Use the solution to Exercise 3.3.

Exercise 3.5   By collecting together group elements according to where they send x, write down the one-one correspondence between the left cosets of $\text{Stab}(x)$ and the elements of $\text{Orb}(x)$ in each of the following cases:

(a) $G = S(\square)$ and $X = \{R, S, T, U\}$, and $x = U$;

(b) $G = S(\triangle)$ and $X = \{R, S, T\}$, the lines of symmetry of the triangle, and $x = T$.

(a)   (b)

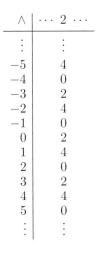

$S(\square)$

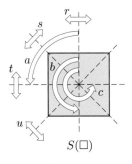

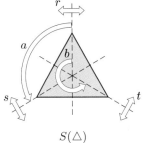

$S(\triangle)$

Hint: The solutions to Exercise 3.1(b) and (c) may be helpful.
For part (a), the action table for this action on page 10 may also be helpful.

We now use the one-one correspondence in the corollary to Theorem 3.2 to prove the Orbit–Stabiliser Theorem.

Theorem 3.1   Orbit–Stabiliser Theorem

Let a finite group G act on a set X. Then, for each $x \in X$,

$$|\text{Orb}(x)| \times |\text{Stab}(x)| = |G|.$$

Proof   Let $x \in X$. There is a one-one correspondence between the left cosets of $\text{Stab}(x)$ and the elements of $\text{Orb}(x)$, so

the number of left cosets of $\text{Stab}(x)$ = the number of elements in $\text{Orb}(x)$.

But the number of left cosets of $\text{Stab}(x)$ is equal to $|G|/|\text{Stab}(x)|$, so

$$|G|/|\text{Stab}(x)| = |\text{Orb}(x)|,$$

and hence

$$|\text{Orb}(x)| \times |\text{Stab}(x)| = |G|. \quad \blacksquare$$

3.3 Groups acting on groups

We conclude this section by looking at some examples of group actions of a group G on a set X, where X is the group G itself, or some set closely related to G. Such actions have important applications in group theory; for example, *conjugation* is a group action of a group on itself, as we see next.

Example 3.1   Let $(G, \circ)$ be a group. Show that

$$g \wedge x = g \circ x \circ g^{-1}, \quad \text{for } g, x \in G,$$

defines a group action of G on itself.

Solution   We show that axioms GA1, GA2 and GA3 hold.

GA1 CLOSURE   For each $g, x \in G$, the conjugate $g \circ x \circ g^{-1}$ is an element of G, so axiom GA1 holds.

GA2 IDENTITY   Let e be the identity in G; then, for each $x \in G$,

$$e \wedge x = e \circ x \circ e^{-1} = x,$$

so axiom GA2 holds.

GA3 COMPOSITION   Let $g_1, g_2, x \in G$. We must show that

$$g_1 \wedge (g_2 \wedge x) = (g_1 \circ g_2) \wedge x.$$

Now

$$g_1 \wedge (g_2 \wedge x) = g_1 \wedge (g_2 \circ x \circ g_2^{-1})$$
$$= g_1 \circ (g_2 \circ x \circ g_2^{-1}) \circ g_1^{-1} \qquad (3.1)$$

and

$$(g_1 \circ g_2) \wedge x = (g_1 \circ g_2) \circ x \circ (g_1 \circ g_2)^{-1}$$
$$= (g_1 \circ g_2) \circ x \circ (g_2^{-1} \circ g_1^{-1}). \qquad (3.2)$$

Recall that $(g_1 \circ g_2)^{-1} = g_2^{-1} \circ g_1^{-1}$. (Unit GTA1, Property 4.4.)

Expressions (3.1) and (3.2) are the same, by associativity in G, so axiom GA3 holds.

Hence $\wedge$ satisfies axioms GA1, GA2 and GA3, so it is a group action. $\quad \blacksquare$

In Example 3.1 we included the binary operation $\circ$ of the group G to highlight the distinction between it and the group action $\wedge$. In the rest of this section, we omit the symbol for the binary operation of a general group G.

Exercise 3.6 Let G be a group. Which of the following define a group action of G on itself? In each case, either show that the three group action axioms hold, or give a counter-example to show that one of them fails.

(a) $g \wedge x = gx$　　(b) $g \wedge x = xg$　　(c) $g \wedge x = xg^{-1}$

Each of the examples of group actions so far in this subsection involves the action of a group on itself. The next group action is slightly different. If G is a group and H is a subgroup of G, then we can define an action of the group H on the set G by

$$h \wedge g = hg, \quad \text{for } h \in H \text{ and } g \in G. \tag{3.3}$$

For example, if G is the cyclic group of order 6 generated by x,

$$G = \{e, x, x^2, x^3, x^4, x^5\},$$

and H is the subgroup $\{e, x^2, x^4\}$, then we get the following action table.

		G					
	$\wedge$	e	x	x^2	x^3	x^4	x^5
	e	e	x	x^2	x^3	x^4	x^5
H	x^2	x^2	x^3	x^4	x^5	e	x
	x^4	x^4	x^5	e	x	x^2	x^3

In the next exercise you are asked to confirm that equation (3.3) always gives a group action.

Exercise 3.7 Let H be a subgroup of a group G. Show that

$$h \wedge g = hg, \quad \text{for } h \in H \text{ and } g \in G,$$

defines a group action of H on G.

We now revisit some topics in group theory in the light of group actions.

Lagrange's Theorem

Let us investigate the orbits of the group action in Exercise 3.7. For any $g \in G$,

$$\mathrm{Orb}(g) = \{h \wedge g : h \in H\} = \{hg : h \in H\}.$$

This is just the right coset Hg, the set obtained by multiplying each element of the subgroup H on the right by g. It follows that the *orbits of this group action* are precisely the *right cosets of H in G*. So the partition of a group G into right cosets of a subgroup H is a special case of the partition of a group into orbits of a group action.

Since the key part of our proof of Lagrange's Theorem is the proof that the cosets of a subgroup in a group form a partition of the group, this tells us that essentially Lagrange's Theorem is a special case of the fact that the orbits of a group action form a partition of the set on which the group acts.

Conjugacy classes

Our next application of group actions to group theory concerns conjugacy. Recall that any group splits into *conjugacy classes*: elements in the same conjugacy class are conjugate to each other, and elements in different classes are not conjugate to each other. For example, the group S_4 has five conjugacy classes, corresponding to the five different cycle structures in S_4:

$\{e\},$

$\{(1\ 2), (1\ 3), (1\ 4), (2\ 3), (2\ 4), (3\ 4)\},$

$\{(1\ 2\ 3), (1\ 2\ 4), (1\ 3\ 2), (1\ 3\ 4), (1\ 4\ 2), (1\ 4\ 3), (2\ 3\ 4), (2\ 4\ 3)\},$

$\{(1\ 2)(3\ 4), (1\ 3)(2\ 4), (1\ 4)(2\ 3)\},$

$\{(1\ 2\ 3\ 4), (1\ 2\ 4\ 3), (1\ 3\ 2\ 4), (1\ 3\ 4\ 2), (1\ 4\ 2\ 3), (1\ 4\ 3\ 2)\}.$

Notice that these classes have 1, 6, 8, 3 and 6 elements respectively, and that each of these numbers divides 24, the order of the group S_4. In the next exercise you are asked to investigate whether the conjugacy classes of some other groups have a similar property.

Exercise 3.8 For each of the following groups, verify that the number of elements in each conjugacy class divides the order of the group.

(a) The cyclic group $\mathbb{Z}_n$

(b) $S(\triangle)$

(c) $S(\square)$

The conjugacy classes of $S(\triangle)$ and $S(\square)$ are given in the Handbook entry for Unit GTB

The result illustrated in Exercise 3.8 holds in *any* finite group. We can prove this by considering the group action of G on itself given by

$$g \wedge x = gxg^{-1}, \quad \text{for } g, x \in G.$$

See Example 3.1.

What are the orbits of this action? For any $x \in G$, we have

$$\begin{aligned} \mathrm{Orb}(x) &= \{g \wedge x : g \in G\} \\ &= \{gxg^{-1} : g \in G\}. \end{aligned}$$

This is just the conjugacy class of x. Thus the *orbits of this group action* are precisely the *conjugacy classes of the group G*. The corollary to the Orbit–Stabiliser Theorem tells us that, for a finite group, the size of any orbit divides the order of the group, so we have the following important result.

Theorem 3.3 In any finite group G, the number of elements in any conjugacy class divides the order of G.

Homomorphisms

Our final application of group actions in this section concerns homomorphisms.

Exercise 3.9 Let $\phi : (G, \circ) \longrightarrow (H, *)$ be a homomorphism. Show that

$$g \wedge h = \phi(g) * h, \quad \text{for } g \in G \text{ and } h \in H,$$

defines a group action of the group G on the set H.

Notice that it is the binary operation of the group H that used in the definition of $\wedge$.

Let us find the orbit and stabiliser of e_H, the identity element of H, under the group action of Exercise 3.9. We have

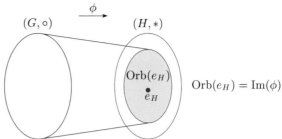

$$\begin{aligned}
\mathrm{Orb}(e_H) &= \{g \wedge e_H : g \in G\} \\
&= \{\phi(g) * e_H : g \in G\} \\
&= \{\phi(g) : g \in G\} = \phi(G) = \mathrm{Im}(\phi)\}
\end{aligned}$$

$\mathrm{Orb}(e_H) = \mathrm{Im}(\phi)$

and

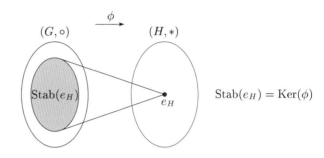

$$\begin{aligned}
\mathrm{Stab}\,(e_H) &= \{g \in G : g \wedge e_H = e_H\} \\
&= \{g \in G : \phi(g) * e_H = e_H\} \\
&= \{g \in G : \phi(g) = e_H\} = \mathrm{Ker}(\phi).\}
\end{aligned}$$

$\mathrm{Stab}(e_H) = \mathrm{Ker}(\phi)$

If G is a finite group, then the Orbit–Stabiliser Theorem tells us that

$$|\mathrm{Orb}(e_H)| \times |\mathrm{Stab}(e_H)| = |G|.$$

For the above group action, this becomes

$$|\mathrm{Im}(\phi)| \times |\mathrm{Ker}(\phi)| = |G|.$$

This is Theorem 4.3 in Unit GTB2, which we deduced from the Correspondence Theorem for homomorphisms. Thus this theorem is a special case of the Orbit–Stabiliser Theorem.

Group actions can be used to prove many further results in group theory. The above discussion illustrates the power of this approach.

Further exercises

Exercise 3.10   Consider the natural group action of S_4 on $X = \{1, 2, 3, 4\}$. For each $x \in X$, determine $|\mathrm{Orb}(x)|$ and $|\mathrm{Stab}(x)|$, and verify that

$$|\mathrm{Orb}(x)| \times |\mathrm{Stab}(x)| = |S_4|.$$

Exercise 3.11   Show that, in general,

$$g \wedge x = g^{-1}xg, \quad \text{for } g, x \in G,$$

does *not* define a group action of a group G on itself.

In Example 3.1 we discovered that, for any group G,

$$g \wedge x = gxg^{-1}, \quad \text{for } g, x \in G,$$

defines a group action of G on itself.

Exercise 3.12   Let G be a group and H be a subgroup of G. Let X be the set of all left cosets of H in G.

(a) Show that

$$g \wedge xH = (gx)H, \quad \text{for } g \in G \text{ and } xH \in X,$$

defines a group action of G on X.

(b) Determine $\mathrm{Orb}(H)$ and $\mathrm{Stab}(H)$ for this action.

(c) Which result can you deduce by applying the Orbit–Stabiliser Theorem (Theorem 3.1) to your answers to part (b)?

Exercise 3.13   Let G be the symmetry group of the cube (which has order 48).

(a) Let G act on the set of six faces of the cube in the natural way. Explain why this action has only one orbit, and deduce that there are eight symmetries of the cube that send a given face to itself.

(b) How many symmetries of the cube:

(i) fix a given vertex?

(ii) send a given edge to itself?

(iii) send a given diagonal to itself?

(iv) send a given diagonal of a face to itself?

A face can be sent to itself without each point being fixed.

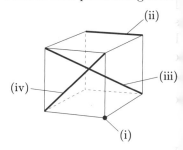

4   The Counting Theorem

After working through this section, you should be able to:

(a) explain what is meant by the *fixed set* of an element of G, when a group G acts on a set X;

(b) determine fixed sets for a given group action;

(c) understand the *Counting Theorem*;

(d) use the Counting Theorem to solve counting problems where a group of symmetries is involved.

4.1 Counting problems

In this section we concentrate on three counting problems. The first of these is as follows.

Question 1   How many different rings can you make using eleven beads, given beads of three different colours?

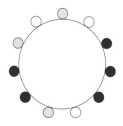

If the ring cannot be rotated or turned over, so that each of the eleven beads is in a fixed position, then this problem can be solved by using the following rule.

Multiplication Rule   If object 1 can be coloured with n_1 colours, object 2 can be coloured with n_2 colours, ..., object k can be coloured with n_k colours, then the number of ways of colouring all k objects is $n_1 n_2 \ldots n_k$.

For example, if there are two objects and each can be coloured from a choice of three colours, then there are $3 \times 3 = 9$ ways of colouring the pair of objects:

However, it is natural to regard some of these colourings as being the same; for example, the second and the fourth colourings each consist of a black object and a white object. In fact, there are just 6 different colourings, if we ignore the order in which the objects are listed.

The situation with the rings is similar. Each of the eleven beads can be coloured in three ways, so there are

$$3 \times 3 \times 3 \times \cdots \times 3 = 3^{11}$$

ways of colouring them. But we shall regard some of these colourings as the same. If we can move one coloured ring so that it appears exactly the same as another, then we regard them as the same ring. For example, we regard the rings

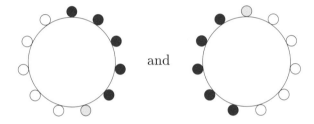

as the same, since either can be rotated to give the other. Similarly, we regard the rings

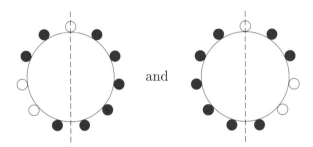

as the same because either can be turned over (or reflected) to give the other. Question 1 can be rephrased as follows.

> How many different rings can be made using eleven beads, given beads of three different colours, if we regard two coloured rings as the same whenever either can be rotated or reflected to give the other?

Exercise 4.1   How many different rings can be made from five beads, given beads of two different colours, if

(a) the positions of the beads are fixed?

(b) rotations and reflections are allowed?

What has all this got to do with group actions? To see this, let X be the set of all 3^{11} coloured rings, and let G be the group of rotations and reflections of such a ring. We can think of the beads as placed at the vertices of a regular 11-gon, so G is essentially the same as the group $S(11\text{-}\mathrm{GON})$; it comprises 11 rotations and 11 reflections.

The group G acts on the set X in the natural way: each element rotates, or reflects, any coloured ring to give another coloured ring. Two coloured rings are thus in the same orbit of this action if either can be rotated or reflected into the other. Thus two rings in the same orbit are rings that we regard as the same, so we can rephrase Question 1 as follows.

How many *orbits* are there in this group action of G on X?

The second counting problem is as follows.

Question 2   How many different patterns can you make by colouring the squares of a chessboard either black or white?

The Multiplication Rule gives one immediate answer to this question. A chessboard has 64 squares and each square can be coloured with a choice of two colours, so the total number of coloured chessboards is

$$2 \times 2 \times 2 \times \cdots \times 2 = 2^{64}.$$

Again, there is a more interesting interpretation of the question. We regard two coloured chessboards as the same if one can be rotated to give the other. For example, we regard the chessboards

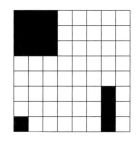 and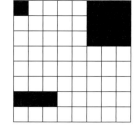

as the same, since a quarter turn anticlockwise takes the first to the second.

With this interpretation, how many differently coloured chessboards are there?

> **Exercise 4.2**   There are $2^4 = 16$ ways of colouring the squares of a 2×2 chessboard black or white. Draw them. How many different colourings of the chessboard are there when we regard two as the same if a rotation takes one to the other?

Question 2 can also be treated as a group action problem. Let X be the set of 2^{64} coloured chessboards and let G be the group $S^+(\square)$ of rotations of the square. This group has order 4, comprising the identity element and the rotations of the square (chessboard) through $\pi/2$, π and $3\pi/2$. The group G acts on the set X in the natural way: each rotation sends a colouring of the chessboard to another colouring. As with Question 1, we regard two colourings as the same precisely when they belong to the same orbit of the action; that is, when one can be rotated to the other. So the answer to Question 2 is again *the number of orbits of the action*.

The third counting problem is as follows.

Question 3   How many different ways can you paint the faces of a cube, given paint of three different colours?

In this case, there are six faces to be coloured, each with a choice of three colours, making 3^6 coloured cubes in all. If we regard two coloured cubes as the same when one can be rotated to give the other, how many different colourings are there?

As with Questions 1 and 2, this can be set up as a group action problem. This time the group G is the group of rotations of the cube and X is the set of all 3^6 coloured cubes. The group G has order 24, consisting of the identity element and the following rotations.

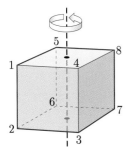

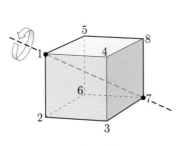

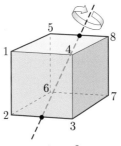

type 1	type 2	type 3
Three axes like this; three rotations about each.	Four axes like this; two rotations about each.	Six axes like this; one rotation about each.

Once again, the required answer is *the number of orbits of the action*.

The programme shows how the number of such orbits can be counted. The following concept is needed.

Definition   Let a group G act on a set X. For $g \in G$, the **fixed set** of g is

$$\mathrm{Fix}(g) = \{x \in X : g \wedge x = x\}.$$

That is, $\mathrm{Fix}(g)$ is the set of elements of X that are fixed by g.

Notice that it is an element of the group G, not an element of the set X, that has a fixed set; this is in contrast to orbits and stabilisers. Fixed sets, like stabilisers, are concerned with elements of G fixing elements of X, but from the opposite point of view. The fixed set of a group element $g \in G$ is the set of all elements of X that are fixed by g, whereas the stabiliser of an element $x \in X$ is the set of all elements of G that fix x. In particular, $\mathrm{Fix}(g)$ is a subset of X, whereas $\mathrm{Stab}(x)$ is a subset of G.

For example, consider the natural action of the group S_3 on the set $\{1, 2, 3\}$, whose action table is reproduced in the margin. For this action,

$\wedge$	1	2	3
e	1	2	3
$(1\ 2\ 3)$	2	3	1
$(1\ 3\ 2)$	3	1	2
$(1\ 2)$	2	1	3
$(1\ 3)$	3	2	1
$(2\ 3)$	1	3	2

$\mathrm{Fix}(e) = \{1, 2, 3\}$,

$\mathrm{Fix}(1\ 2\ 3) = \varnothing$,

$\mathrm{Fix}(1\ 3\ 2) = \varnothing$,

$\mathrm{Fix}(1\ 2) = \{3\}$,

$\mathrm{Fix}(1\ 3) = \{2\}$,

$\mathrm{Fix}(2\ 3) = \{1\}$.

The symbol $\varnothing$ denotes the empty set.

Notice that fixed sets, like stabilisers, can be found by looking through the action table for where elements are fixed; for fixed sets we read across rows, whereas for stabilisers we read down columns.

Exercise 4.3   Consider the group action of $S(\square)$ on the lines of symmetry of the square, whose action table is reproduced in the margin. Find the fixed set of each element of $S(\square)$ for this action.

$\wedge$	R	S	T	U
e	R	S	T	U
a	T	U	R	S
b	R	S	T	U
c	T	U	R	S
r	R	U	T	S
s	T	S	R	U
t	R	U	T	S
u	T	S	R	U

Remark   The fixed point sets that you met in Unit GTB1 are special cases of fixed sets. Recall that if g is a symmetry of a figure F, then the *fixed point set* of g is the set of all points of F that are fixed by g. This is the fixed set of g under the natural group action of the symmetry group $S(F)$ on the set of points of F.

The three counting problems discussed in this subsection are solved in the video programme.

Watch the video programme '57 varieties'.

Video

4.2 Review of the programme

In the programme we pose the three questions introduced in Section 4.1, and we set them up as problems in counting orbits of group actions. Each of the three questions can be answered by applying the following theorem, which is illustrated in the programme.

Theorem 4.1 Counting Theorem

Let $\wedge$ be a group action of a finite group G on a finite set X. Then the number t of orbits of the action is given by the formula

$$t = \frac{1}{|G|} \sum_{g \in G} |\mathrm{Fix}(g)|.$$

We give a proof of the Counting Theorem in Section 4.4.

$|\mathrm{Fix}(g)|$ denotes the number of elements in $\mathrm{Fix}(g)$.

For each of the three questions we use a similar method. To find the number of orbits, we determine the number $|\mathrm{Fix}(g)|$ for each element g in the group, add up all these numbers, and divide the total by $|G|$. The details of the calculations are as follows.

Question 1   For the rings problem, the 22 elements of $S(11\text{-GON})$ act on the 3^{11} coloured rings, as follows.

(a) *The identity e.* This fixes all 3^{11} coloured rings, so $|\mathrm{Fix}(e)| = 3^{11}$.

(b) *Ten rotations through $2\pi/11, 4\pi/11, \ldots, 20\pi/11$.* The only coloured rings fixed by these rotations are the three rings in which all eleven beads are the same colour. Thus $|\mathrm{Fix}(g)| = 3$ for each of these rotations, as there are 3 colours.

(c) *Eleven reflections.* Each reflection fixes one bead on its axis and transposes the other beads in five pairs. For a coloured ring to be fixed by a particular reflection, any two beads transposed as a pair must be the same colour, but beads from different pairs may be different colours. The bead on the axis of the reflection may be any colour. So there are six colour selections to be made (one for the fixed bead and one for each of the five pairs), each from a choice of 3 colours. This makes a total of 3^6 coloured rings, so $|\mathrm{Fix}(g)| = 3^6$ for each reflection.

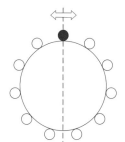

By the Counting Theorem, the number of orbits is

$$\tfrac{1}{22}(3^{11} + (10 \times 3) + (11 \times 3^6)) = 8418.$$

This is the number of differently coloured rings.

Thus the Counting Theorem has reduced a complicated counting problem to a straightforward calculation—such is the power of group theory!

Question 2   For the chessboard problem, the four elements of $S^+(\square)$ act on the 2^{64} coloured chessboards, as follows.

(a) *The identity element.* This fixes all the coloured chessboards, so $|\text{Fix}(e)| = 2^{64}$.

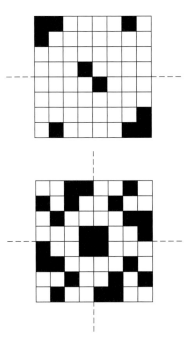

(b) *The rotation through π.* This transposes diagonally opposite squares of the chessboard. So for a coloured chessboard to be fixed by this rotation, each square must be the same colour as its diagonally opposite square. Therefore we can colour the top half of the chessboard in any of the 2^{32} possible ways and this determines the colouring of the lower half of the chessboard. Thus $|\text{Fix}(g)| = 2^{32}$ for this rotation.

(c) *The rotations through $\pi/2$ and $3\pi/2$.* For a coloured chessboard to be fixed by one of these rotations, each square must be the same colour as the three squares onto which it maps under successive quarter turns. Therefore one quarter of the board can be coloured in any of the 2^{16} possible ways, thereby determining the colours of the squares in the other quarters of the board. Thus $|\text{Fix}(g)| = 2^{16}$ for each of these two rotations.

By the Counting Theorem, the number of orbits is

$$\tfrac{1}{4} \times (2^{64} + 2^{32} + (2 \times 2^{16})) = 2^{15} \left(2^{47} + 2^{15} + 1\right),$$

which is an enormous number with 19 digits.

This is the number of differently coloured chessboards.

Question 3   For the cubes problem, the 24 elements of the group of rotations of the cube act on the 3^6 coloured cubes. The 24 group elements are the identity and the following rotations.

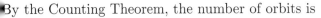

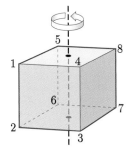

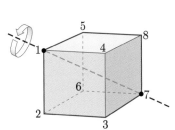

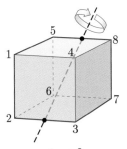

type 1	type 2	type 3
Three axes like this; three rotations about each.	Four axes like this; two rotations about each.	Six axes like this; one rotation about each.

We can split these 24 group elements into five classes, as follows.

(a) *The identity element.* This fixes all the coloured cubes, so $|\text{Fix}(e)| = 3^6$.

(b) *Three rotations of type 1 through π.* Each such rotation fixes the two faces through which the axis of rotation passes, and transposes the other four faces in two pairs. For a coloured cube to be fixed, the two faces through which the axis of rotation passes can be coloured arbitrarily, but the two faces in each pair of transposed faces must be coloured the same. Thus there are four colour selections to be made, so $|\text{Fix}(g)| = 3^4$ for each of these rotations.

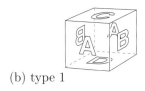

(b) type 1

(c) *Six rotations of type 1 through $\pi/2$ and $3\pi/2$.* Each such rotation fixes the two faces through which the axis passes, and maps the other four faces in a 4-cycle. The two fixed faces can be coloured arbitrarily, but the four remaining faces must be coloured the same. Thus $|\text{Fix}(g)| = 3^3$ for each of these rotations.

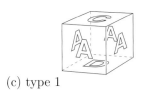

(c) type 1

41

(d) *Eight rotations of type 2.* For each of the two vertices on the axis of rotation, the three adjacent faces are mapped to each other in a 3-cycle and so must be coloured the same. Thus $|\text{Fix}(g)| = 3^2$ for each of these rotations.

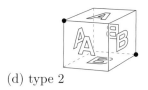

(d) type 2

(e) *Six rotations of type 3.* The faces are transposed in three pairs; the two faces in each pair must be coloured the same. Thus $|\text{Fix}(g)| = 3^3$ for each of these elements.

By the Counting Theorem, the number of orbits is

$$\tfrac{1}{24}(3^6 + (3 \times 3^4) + (6 \times 3^3) + (8 \times 3^2) + (6 \times 3^3)) = 57.$$

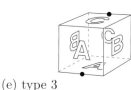

(e) type 3

There are thus 57 varieties of coloured cube.

Post-programme exercises

Exercise 4.4   Use the Counting Theorem to determine how many differently coloured rings of five beads can be made using beads of two different colours, when we regard two rings as the same if a rotation or reflection takes one to the other. (Hence check your answer to Exercise 4.1.)

Exercise 4.5   Use the Counting Theorem to determine how many ways there are of colouring a 2×2 chessboard with two colours, when we regard two chessboards as the same if a rotation takes one to the other. (Hence check your answer to Exercise 4.2.)

Exercise 4.6   How many ways are there of colouring the squares of a 4×4 chessboard with two colours, when we regard two chessboards as the same if a rotation takes one to the other?

Exercise 4.7   Not all the 57 differently coloured cubes of Question 3 (page 38) use all three colours. Use the Counting Theorem to determine how many differently coloured cubes there are when only two colours are available.

Exercise 4.7 illustrates that changing the number of colours in a problem of this type results in only a minor modification of the calculation.

4.3 Proof of the Counting Theorem (optional)

We now prove the theorem used in the video programme.

Theorem 4.1 Counting Theorem

Let $\wedge$ be a group action of a finite group G on a finite set X. Then the number t of orbits of the action is given by the formula

$$t = \frac{1}{|G|} \sum_{g \in G} |\text{Fix}(g)|.$$

Proof   The proof is in three parts.

(a) Let $\mathrm{Orb}(x)$ be any one of the t orbits of the group action, and consider the sum

$$\sum_{y \in \mathrm{Orb}(x)} |\mathrm{Stab}(y)|.$$

By the Orbit–Stabiliser Theorem (Theorem 3.1), we have

$$\sum_{y \in \mathrm{Orb}(x)} |\mathrm{Stab}(y)| = \sum_{y \in \mathrm{Orb}(x)} \frac{|G|}{|\mathrm{Orb}(y)|} = |G| \sum_{y \in \mathrm{Orb}(x)} \frac{1}{|\mathrm{Orb}(y)|}.$$

But if $y \in \mathrm{Orb}(x)$, then $\mathrm{Orb}(y) = \mathrm{Orb}(x)$, so all the terms in the summation are the same, equal to $1/|\mathrm{Orb}(x)|$. There are $|\mathrm{Orb}(x)|$ terms in the summation, so the right-hand side of the above equation becomes

$$|G| \times |\mathrm{Orb}(x)| \times \frac{1}{|\mathrm{Orb}(x)|},$$

which is $|G|$. Thus

$$\sum_{y \in \mathrm{Orb}(x)} |\mathrm{Stab}(y)| = |G|.$$

(b) We now add together the results of part (a) for each of the t orbits. This gives

$$\sum_{t \text{ orbits}} \left(\sum_{y \in \mathrm{Orb}(x)} |\mathrm{Stab}(y)| \right) = t|G|.$$

But if y runs through a particular orbit and then through all orbits, then y must run through *all* the elements of X, since each element of X lies in exactly one orbit. Therefore

$$\sum_{y \in X} |\mathrm{Stab}(y)| = t|G|,$$

so

$$t = \frac{1}{|G|} \sum_{y \in X} |\mathrm{Stab}(y)|.$$

(c) It remains to show that $\sum_{y \in X} |\mathrm{Stab}(y)| = \sum_{g \in G} |\mathrm{Fix}(g)|$.

To do this, we consider the action table and count up the number of cells in the table where $g \wedge y = y$.

$$X$$

$\wedge$		$\cdots$		y		$\cdots$	
$\vdots$				$\vdots$			
g	$\cdots$	✓	✓	$\cdots$ ✓	$\cdots$	✓	$\cdots$
$\vdots$				$\vdots$			
				✓			
				✓			
				$\vdots$			

We place a tick in each cell of the table where an element y of X is fixed by an element g of G.

Now

the number of ticks in the column corresponding to $y \in X$
$= |\text{Stab}(y)|$, the number of group elements stabilising y.

Summing over all the columns, we see that

the total number of ticks in the table is $\displaystyle\sum_{y \in X} |\text{Stab}(y)|$.

But

the number of ticks in the row corresponding to $g \in G$
$= |\text{Fix}(g)|$, the number of elements of X fixed by g.

Summing over all the rows, we see that

the total number of ticks in the table is $\displaystyle\sum_{g \in G} |\text{Fix}(g)|$.

Thus

$$\sum_{y \in X} |\text{Stab}(y)| = \sum_{g \in G} |\text{Fix}(g)|.$$

This completes the proof.   ■

Further exercises

Exercise 4.8

(a) Use the Counting Theorem to determine how many different
rectangular blankets, similar to the one illustrated in the margin, can
be made if each quadrant is to be coloured with one of two colours,
and we regard two blankets as the same if a rotation or reflection takes
one to the other.

(b) Confirm your answer to part (a) by drawing the possibilities.

Exercise 4.9

(a) Use the Counting Theorem to determine how many different triangular
window stickers, similar to the one illustrated in the margin, can be
made if each (equilateral) triangle region is to be coloured with one of
two colours, and we regard two stickers as the same if a rotation or a
reflection takes one to the other.

(b) Confirm your answer to part (a) by drawing the possibilities.

(c) How many different window stickers can be made if three colours can
be used?

Solutions to the exercises

1.1 (a) The answers can be read from the appropriate rows of the action table. From the row labelled (1 2), we see that (1 2) sends 2 to 1, so

$$(1\ 2) \wedge 2 = 1.$$

Similarly,

$$(1\ 2) \wedge 3 = 3,$$
$$(1\ 3\ 2) \wedge 1 = 3,$$
$$(1\ 3) \wedge 2 = 2.$$

(b) The identity element sends each of the three symbols to itself; that is, it *fixes* each symbol, so

$$e \wedge x = x, \quad \text{for } x = 1, 2 \text{ and } 3.$$

1.2 (a) From the action table, we obtain

$$r \wedge 2 = 3, \quad c \wedge 3 = 2, \quad b \wedge 1 = 3, \quad u \wedge 4 = 4.$$

(b) From the row labelled e in the table, we see that

$$e \wedge 1 = 1, \quad e \wedge 2 = 2, \quad e \wedge 3 = 3, \quad e \wedge 4 = 4.$$

(c) From the table, we find that

$$u \wedge 2 = 2,$$

so

$$b \wedge (u \wedge 2) = b \wedge 2 = 4$$

and

$$(b \circ u) \wedge 2 = s \wedge 2 = 4.$$

Thus

$$b \wedge (u \wedge 2) = (b \circ u) \wedge 2.$$

Similarly,

$$s \wedge (c \wedge 4) = s \wedge 3 = 3$$

and

$$(s \circ c) \wedge 4 = t \wedge 4 = 3.$$

Thus

$$s \wedge (c \wedge 4) = (s \circ c) \wedge 4.$$

The general rule illustrated here is that, for all $g_1, g_2 \in S(\square)$ and all $x \in \{1, 2, 3, 4\}$,

$$g_1 \wedge (g_2 \wedge x) = (g_1 \circ g_2) \wedge x.$$

(We show later than the action of $S(\square)$ on $\{1, 2, 3, 4\}$ satisfies this rule.)

1.3 The complete table is as follows.

$\wedge$	R	S	T	U
e	R	S	T	U
a	T	U	R	S
b	R	S	T	U
c	T	U	R	S
r	R	U	T	S
s	T	S	R	U
t	R	U	T	S
u	T	S	R	U

1.4 By axiom GA2, we have

$$e \wedge b = b, \quad e \wedge c = c, \quad e \wedge d = d;$$

this enables us to complete the first row of the table. The entries in the other highlighted cells can be found as follows.

$$(1\ 2)(3\ 4) \wedge c$$
$$= (1\ 2)(3\ 4) \wedge ((1\ 3)(2\ 4) \wedge a) \quad \text{(from the table)}$$
$$= ((1\ 2)(3\ 4) \circ (1\ 3)(2\ 4)) \wedge a \quad \text{(axiom GA3)}$$
$$= (1\ 4)(2\ 3) \wedge a \quad \text{(compose in } K)$$
$$= b \quad \text{(from the table)}.$$

$$(1\ 3)(2\ 4) \wedge b$$
$$= (1\ 3)(2\ 4) \wedge ((1\ 4)(2\ 3) \wedge a) \quad \text{(from the table)}$$
$$= ((1\ 3)(2\ 4) \circ (1\ 4)(2\ 3)) \wedge a \quad \text{(axiom GA3)}$$
$$= (1\ 2)(3\ 4) \wedge a \quad \text{(compose in } K)$$
$$= d \quad \text{(from the table)}.$$

$$(1\ 4)(2\ 3) \wedge d$$
$$= (1\ 4)(2\ 3) \wedge ((1\ 2)(3\ 4) \wedge a) \quad \text{(from the table)}$$
$$= ((1\ 4)(2\ 3) \circ (1\ 2)(3\ 4)) \wedge a \quad \text{(axiom GA3)}$$
$$= (1\ 3)(2\ 4) \wedge a \quad \text{(compose in } K)$$
$$= c \quad \text{(from the table)}.$$

Remark All the entries in the last three rows of the table can be found in a similar way. The complete table is as follows.

$\wedge$	a	b	c	d
e	a	b	c	d
$(1\ 2)(3\ 4)$	d	c	b	a
$(1\ 3)(2\ 4)$	c	d	a	b
$(1\ 4)(2\ 3)$	b	a	d	c

1.5 The action tables are as follows.

(a)

$\wedge$	1	2	3
e	1	2	3
a	2	3	1
b	3	1	2
r	1	3	2
s	3	2	1
t	2	1	3

$S(\triangle)$ acting on $\{1, 2, 3\}$

(b)

$\wedge$	A	B	C	D
e	A	B	C	D
a	B	C	D	A
b	C	D	A	B
c	D	A	B	C
r	A	D	C	B
s	B	A	D	C
t	C	B	A	D
u	D	C	B	A

$S(\square)$ acting on $\{A, B, C, D\}$

We show that axioms GA1, GA2 and GA3 hold.

GA1 In each case, the only elements occurring in the body of the table are elements of the appropriate set, so axiom GA1 holds.

GA2 In each case, the first row of the table confirms that the identity element of the group fixes each element of the set, so axiom GA2 holds.

GA3 In each case, we have a group of functions with the binary operation of function composition, and the group action is

$$g \wedge x = g(x),$$

for all group elements g and all elements x of the set being acted on. Hence axiom GA3 is just the rule for composition of functions:

$$g_1(g_2(x)) = (g_1 \circ g_2)(x),$$

for all group elements g_1, g_2 and all elements x of the set acted on. This is true, so axiom GA3 holds.

Hence, in each case, $\wedge$ satisfies axioms GA1, GA2 and GA3, so it is a group action. Thus both tables are action tables of group actions.

1.6 (a) We can take $\{a, b, c, d\}$ to be the set of vertices of the rectangle, with a an alternative label for 1, b for 2, c for 3 and d for 4. Then the given table is the action table for the natural action of K on the set of vertices of the rectangle.

Alternatively, we can take $\{a, b, c, d\}$ to be the set of four quadrants of the rectangle, as illustrated.

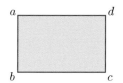

The given table is then the action table for the natural action of K on the set of quadrants.

(b) We are looking for a set of four features $\{a, b, c, d\}$ of the rectangle, such that

a and c are fixed by all four symmetries of the rectangle;

b and d are fixed by the identity and the rotation through π, but are transposed by the two reflections.

One possibility is to take b and d to be the diagonals of the rectangle, and a and c to be the horizontal and vertical lines of symmetry, as illustrated.

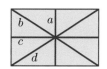

1.7 Part of the action table is as follows.

$\wedge$	0	1	2	3
$\vdots$	$\vdots$	$\vdots$	$\vdots$	$\vdots$
-1	1	2	3	0
0	0	1	2	3
1	3	0	1	2
2	2	3	0	1
3	1	2	3	0
4	0	1	2	3
$\vdots$	$\vdots$	$\vdots$	$\vdots$	$\vdots$

We show that axioms GA1, GA2 and GA3 hold.

GA1 The value modulo 4 of any integer is an element of the set $X = \{0, 1, 2, 3\}$, so axiom GA1 holds.

GA2 For each $x \in X$, $0 \wedge x = x$, so axiom GA2 holds.

GA3 Let m and n be any elements of $\mathbb{Z}$ and let x be any element of X. We have

$$m \wedge (n \wedge x)$$
$$= m \wedge (3n + x) \pmod 4$$
$$= (3m + 3n + x) \pmod 4$$

and

$$(m + n) \wedge x$$
$$= (3(m + n) + x) \pmod 4$$
$$= (3m + 3n + x) \pmod 4.$$

These two expressions are the same, so axiom GA3 holds.

Hence $\wedge$ satisfies axioms GA1, GA2 and GA3, so it is a group action.

1.8 (a)

$g \wedge A$ for $g = e, b$

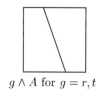

$g \wedge A$ for $g = a, c$

$g \wedge A$ for $g = s, u$

$g \wedge A$ for $g = r, t$

(b)

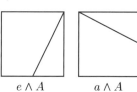

$e \wedge A \qquad a \wedge A \qquad b \wedge A$

$c \wedge A$

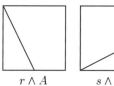

$r \wedge A \qquad s \wedge A$

$t \wedge A$

$u \wedge A$

(c)

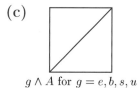

$g \wedge A$ for $g = e, b, s, u$

$g \wedge A$ for $g = a, c, r, t$

(d)

$g \wedge A$ for each $g \in S(\square)$

1.9 (a) We check axioms GA1, GA2 and GA3.

GA1 The element $(ax, y) \in \mathbb{R}^2$, for all real numbers a, x and y, so axiom GA1 holds.

GA2 The identity element of G is $\begin{pmatrix} 1 & 0 \\ 0 & 1 \end{pmatrix}$ and

$$\begin{pmatrix} 1 & 0 \\ 0 & 1 \end{pmatrix} \wedge (x, y) = (x, y),$$

for each $(x, y) \in \mathbb{R}^2$, so axiom GA2 holds.

GA3 Let $\begin{pmatrix} a & b \\ 0 & 1 \end{pmatrix}, \begin{pmatrix} c & d \\ 0 & 1 \end{pmatrix} \in G$ and $(x, y) \in \mathbb{R}^2$.
We have to show that

$$\begin{pmatrix} a & b \\ 0 & 1 \end{pmatrix} \wedge \left(\begin{pmatrix} c & d \\ 0 & 1 \end{pmatrix} \wedge (x, y) \right)$$
$$= \left(\begin{pmatrix} a & b \\ 0 & 1 \end{pmatrix} \begin{pmatrix} c & d \\ 0 & 1 \end{pmatrix} \right) \wedge (x, y).$$

We have

$$\begin{pmatrix} a & b \\ 0 & 1 \end{pmatrix} \wedge \left(\begin{pmatrix} c & d \\ 0 & 1 \end{pmatrix} \wedge (x, y) \right)$$
$$= \begin{pmatrix} a & b \\ 0 & 1 \end{pmatrix} \wedge (cx, y)$$
$$= (acx, y) \tag{S.1}$$

and

$$\left(\begin{pmatrix} a & b \\ 0 & 1 \end{pmatrix} \begin{pmatrix} c & d \\ 0 & 1 \end{pmatrix} \right) \wedge (x, y)$$
$$= \begin{pmatrix} ac & ad+b \\ 0 & 1 \end{pmatrix} \wedge (x, y)$$
$$= (acx, y). \tag{S.2}$$

Expressions (S.1) and (S.2) are the same, so axiom GA3 holds.

Hence $\wedge$ satisfies axioms GA1, GA2 and GA3, so it is a group action.

(b) Axiom GA2 does not hold because, for example,

$$\begin{pmatrix} 1 & 0 \\ 0 & 1 \end{pmatrix} \wedge (2, 2) = (2, 0) \neq (2, 2).$$

Thus $\wedge$ is not a group action.

(Axiom GA3 does not hold either, but axiom GA1 holds.)

(c) Axioms GA1 and GA2 hold, but axiom GA3 does not hold.

For axiom GA3 to hold, we require that, for all $\begin{pmatrix} a & b \\ 0 & 1 \end{pmatrix}, \begin{pmatrix} c & d \\ 0 & 1 \end{pmatrix} \in G$ and all $(x, y) \in \mathbb{R}^2$,

$$\begin{pmatrix} a & b \\ 0 & 1 \end{pmatrix} \wedge \left(\begin{pmatrix} c & d \\ 0 & 1 \end{pmatrix} \wedge (x, y) \right)$$
$$= \left(\begin{pmatrix} a & b \\ 0 & 1 \end{pmatrix} \begin{pmatrix} c & d \\ 0 & 1 \end{pmatrix} \right) \wedge (x, y).$$

The left-hand side of this equation is equal to

$$\begin{pmatrix} a & b \\ 0 & 1 \end{pmatrix} \wedge (cx, dy + y)$$
$$= (acx, b(dy + y) + dy + y)$$
$$= (acx, bdy + by + dy + y) \tag{S.3}$$

and the right-hand side is equal to

$$\begin{pmatrix} ac & ad+b \\ 0 & 1 \end{pmatrix} \wedge (x, y)$$
$$= (acx, (ad + b)y + y)$$
$$= (acx, ady + by + y). \tag{S.4}$$

Expressions (S.3) and (S.4) are equal only if

$$bdy + by + dy + y = ady + by + y;$$

that is, only if

$$bdy + dy = ady.$$

This equation is not true in general. For instance, taking $a = b = d = y = 1$, it gives $2 = 1$.

Thus $\wedge$ is not a group action.

1.10

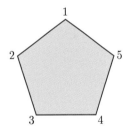

The action table for $S(\Circle)$ acting on $\{1, 2, 3, 4, 5\}$ is as follows.

$\wedge$	1	2	3	4	5
e	1	2	3	4	5
(1 2 3 4 5)	2	3	4	5	1
(1 3 5 2 4)	3	4	5	1	2
(1 4 2 5 3)	4	5	1	2	3
(1 5 4 3 2)	5	1	2	3	4
(1 2)(3 5)	2	1	5	4	3
(1 3)(4 5)	3	2	1	5	4
(1 4)(2 3)	4	3	2	1	5
(1 5)(2 4)	5	4	3	2	1
(2 5)(3 4)	1	5	4	3	2

1.11 (a) The action table is as follows.

$\wedge$	0	1	2
0	0	1	2
1	1	2	0
2	2	0	1
3	0	1	2
4	1	2	0
5	2	0	1

This defines a group action. We show that axioms GA1, GA2 and GA3 hold.

GA1 The body of the table contains only elements from X, so axiom GA1 holds.

GA2 The identity element in $\mathbb{Z}_6$ is 0, and from the row labelled 0 we see that

$$0 \wedge x = x, \quad \text{for } x = 0, 1, 2,$$

so axiom GA2 holds.

GA3 Let $g, h \in G$ and let $x \in X$. We have

$$g \wedge (h \wedge x) = g \wedge (h +_3 x)$$
$$= g +_3 (h +_3 x) \qquad (\text{S.5})$$

and

$$(g +_6 h) \wedge x = (g +_6 h) +_3 x. \qquad (\text{S.6})$$

Now

$$g +_3 (h +_3 x) = (g +_6 h) +_3 x,$$

since each side is equal to the remainder when $g + h + x$ is divided by 3. Thus expressions (S.5) and (S.6) are equal, so axiom GA3 holds.

Hence $\wedge$ satisfies axioms GA1, GA2 and GA3, so it is a group action.

(b) The action table is as follows.

$\wedge$	0	1	2	3
0	0	1	2	3
1	1	2	3	0
2	2	3	0	1
3	3	0	1	2
4	0	1	2	3
5	1	2	3	0

This is not a group action. Axioms GA1 and GA2 hold, but axiom GA3 fails. For example,

$$4 \wedge (3 \wedge 1) = 4 \wedge 0 = 0$$

but

$$(4 +_6 3) \wedge 1 = 1 \wedge 1 = 2,$$

so

$$4 \wedge (3 \wedge 1) \neq (4 +_6 3) \wedge 1.$$

1.12 We show that axioms GA1, GA2 and GA3 hold.

GA1 Let $g \in S_4$ and let $\{i, j\} \in X$; then $i \neq j$ and

$$g \wedge \{i, j\} = \{g(i), g(j)\}.$$

Since g is a permutation, $g(i)$ and $g(j)$ are distinct symbols, so $\{g(i), g(j)\} \in X$. Thus axiom GA1 holds.

GA2 For each $\{i, j\} \in X$,

$$e \wedge \{i, j\} = \{e(i), e(j)\} = \{i, j\},$$

so axiom GA2 holds.

GA3 Let $g, h \in S_4$ and $\{i, j\} \in X$; then

$$g \wedge (h \wedge \{i, j\}) = g \wedge \{h(i), h(j)\}$$
$$= \{g(h(i)), g(h(j))\}$$
$$= \{(g \circ h)(i), (g \circ h)(j)\}$$
$$= (g \circ h) \wedge \{i, j\}.$$

Thus axiom GA3 holds.

Hence $\wedge$ satisfies axioms GA1, GA2 and GA3, so it a group action.

1.13 We show that axioms GA1, GA2 and GA3 hold.

GA1 For all $r \in \mathbb{R}$ and all $(x, y) \in \mathbb{R}^2$,

$$r \wedge (x, y) = (x, y + r) \in \mathbb{R}^2,$$

so axiom GA1 holds.

GA2 The identity in $\mathbb{R}$ is 0 and, for each $(x, y) \in \mathbb{R}$

$$0 \wedge (x, y) = (x, y + 0) = (x, y),$$

so axiom GA2 holds.

GA3 For all $r, s \in \mathbb{R}$ and all $(x, y) \in \mathbb{R}^2$,

$$r \wedge (s \wedge (x, y)) = r \wedge (x, y + s)$$
$$= (x, (y + s) + r) \qquad (\text{S.7})$$

and

$$(r + s) \wedge (x, y) = (x, y + (r + s)). \qquad (\text{S.8})$$

Expressions (S.7) and (S.8) are equal, by the associative and commutative properties of the addition of real numbers, so axiom GA3 holds

Hence $\wedge$ satisfies axioms GA1, GA2 and GA3, so it a group action.

2.1 Here

$$\text{Orb}(x, y) = \{(ax, by) : a, b \in \mathbb{R}^+\},$$

so

$\text{Orb}(1, 0) = \{(a, 0) : a \in \mathbb{R}^+\},$ the positive x-axis;

$\text{Orb}(0, 1) = \{(0, b) : b \in \mathbb{R}^+\},$ the positive y-axis;

$\text{Orb}(0, -1) = \{(0, -b) : b \in \mathbb{R}^+\},$ the negative y-axis

$\text{Orb}(1, 1) = \{(a, b) : a, b \in \mathbb{R}^+\},$ the first quadrant

$\text{Orb}(-1, 1) = \{(-a, b) : a, b \in \mathbb{R}^+\},$ the second quadrant

$\text{Orb}(-1, -1) = \{(-a, -b) : a, b \in \mathbb{R}^+\},$ the third quadrant

There are nine orbits in all: the origin, the four semi-axes and the four quadrants, as illustrated in the following diagram.

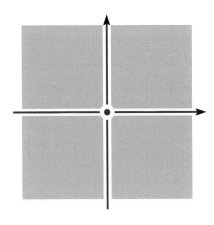

2.2 Here

$$\text{Orb}(x, y) = \left\{ \begin{pmatrix} a & b \\ 0 & 1 \end{pmatrix} \wedge (x, y) : a, b \in \mathbb{R},\ a \neq 0 \right\}$$
$$= \{(ax, y) : a, b \in \mathbb{R},\ a \neq 0\}$$
$$= \{(ax, y) : a \in \mathbb{R}^*\}.$$

When $x = 0$, we have

$$\text{Orb}(0, y) = \{(a0, y) : a \in \mathbb{R}^*\} = \{(0, y)\},$$

so each point on the y-axis lies in an orbit containing itself alone. For example, $\text{Orb}(0, 2) = \{(0, 2)\}$.

When $x \neq 0$,

$$\text{Orb}(x, y) = \{(ax, y) : a \in \mathbb{R}^*\},$$

from above, and this is the set of all points on the line through the point (x, y) parallel to the x-axis, except for the point $(0, y)$. There is one such orbit for each value of y. For example, $\text{Orb}(1, 2)$ is the line $y = 2$ excluding the point $(0, 2)$.

Thus there are infinitely many orbits: each point on the y-axis is a one-point orbit and each horizontal line (excluding the point on the y-axis) is an orbit.

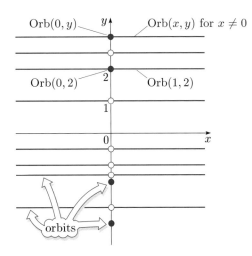

2.3 We follow Strategy 2.1 (see Frame 10).

(a) First we find $\text{Orb}(x)$ for any $x \in X$, say $x = 0$:

$$\text{Orb}(0) = \{g \wedge 0 : g \in \mathbb{Z}\}$$
$$= \{(2g + 0) \pmod 6 : g \in \mathbb{Z}\}$$
$$= \{0, 2, 4\}.$$

Next we choose an element not in the orbit already found, say $x = 1$:

$$\text{Orb}(1) = \{g \wedge 1 : g \in \mathbb{Z}\}$$
$$= \{(2g + 1) \pmod 6 : g \in \mathbb{Z}\}$$
$$= \{1, 3, 5\}.$$

Every element of X has now been assigned to an orbit. Thus there are two orbits,

$$\{0, 2, 4\} \text{ and } \{1, 3, 5\}.$$

(b)

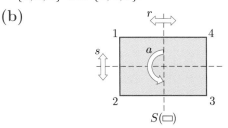

First we find $\text{Orb}(x)$ for any $x \in X$, say $x = 1$:

$$\text{Orb}(1) = \{e \wedge 1, s \wedge 1, a \wedge 1, r \wedge 1\}$$
$$= \{1, 2, 3, 4\}.$$

Every element of X has now been assigned to an orbit. Thus there is just one orbit—X itself.

2.4 (a) For the group action in Exercise 2.1,

$$\text{Stab}(x, y)$$
$$= \left\{ \begin{pmatrix} a & 0 \\ 0 & b \end{pmatrix} \in G : \begin{pmatrix} a & 0 \\ 0 & b \end{pmatrix} \wedge (x, y) = (x, y) \right\}$$
$$= \left\{ \begin{pmatrix} a & 0 \\ 0 & b \end{pmatrix} : (ax, by) = (x, y),\ a, b \in \mathbb{R}^+ \right\}. \quad \text{(S.9)}$$

We consider separately the four possible cases depending on whether or not x and/or y is 0.

First we find the stabiliser of the origin. Taking $x = y = 0$ in equation (S.9) gives

$$\text{Stab}(0, 0)$$
$$= \left\{ \begin{pmatrix} a & 0 \\ 0 & b \end{pmatrix} : (0, 0) = (0, 0),\ a, b \in \mathbb{R}^+ \right\}.$$

The condition $(0, 0) = (0, 0)$ is true for all a and b in $\mathbb{R}^+$, so

$$\text{Stab}(0, 0) = \left\{ \begin{pmatrix} a & 0 \\ 0 & b \end{pmatrix} : a, b \in \mathbb{R}^+ \right\} = G.$$

Next we find the stabilisers of points on the x-axis, except the origin. If $x \neq 0$ and $y = 0$, equation (S.9) becomes

$$\text{Stab}(x, 0)$$
$$= \left\{ \begin{pmatrix} a & 0 \\ 0 & b \end{pmatrix} : (ax, b0) = (x, 0),\ a, b \in \mathbb{R}^+ \right\},$$

so $a = 1$ and b can be any positive real number. Thus

$$\text{Stab}(x, 0) = \left\{ \begin{pmatrix} 1 & 0 \\ 0 & b \end{pmatrix} : b \in \mathbb{R}^+ \right\}.$$

Next we find the stabilisers of points on the y-axis, except the origin. If $x = 0$ and $y \neq 0$, equation (S.9) becomes

$$\text{Stab}(0, y)$$
$$= \left\{ \begin{pmatrix} a & 0 \\ 0 & b \end{pmatrix} : (a0, by) = (0, y),\ a, b \in \mathbb{R}^+ \right\},$$

so a can be any positive real number and $b = 1$. Thus

$$\text{Stab}(0, y) = \left\{ \begin{pmatrix} a & 0 \\ 0 & 1 \end{pmatrix} : a \in \mathbb{R}^+ \right\}.$$

Finally, we find the stabilisers of points off the axes. If $x \neq 0$ and $y \neq 0$, then it follows from equation (S.9) that $a = b = 1$. Thus

$$\text{Stab}(x, y) = \left\{ \begin{pmatrix} 1 & 0 \\ 0 & 1 \end{pmatrix} \right\}.$$

So any point off the axes is stabilised by the identity element alone.

(b) For the group action in Exercise 2.2,

$$\text{Stab}(x, y)$$
$$= \left\{ \begin{pmatrix} a & b \\ 0 & 1 \end{pmatrix} \in G : \begin{pmatrix} a & b \\ 0 & 1 \end{pmatrix} \wedge (x, y) = (x, y) \right\}$$
$$= \left\{ \begin{pmatrix} a & b \\ 0 & 1 \end{pmatrix} : (ax, y) = (x, y),\ a, b \in \mathbb{R},\ a \neq 0 \right\}.$$

We consider separately the points on the y-axis and the points off the y-axis.

If $x = 0$, then

$$\text{Stab}(0, y)$$
$$= \left\{ \begin{pmatrix} a & b \\ 0 & 1 \end{pmatrix} : (0, y) = (0, y),\ a, b \in \mathbb{R}, a \neq 0 \right\}$$
$$= G.$$

Thus each point on the y-axis is stabilised by the whole group.

If $x \neq 0$, then we must have $a = 1$, so

$$\text{Stab}(x, y) = \left\{ \begin{pmatrix} 1 & b \\ 0 & 1 \end{pmatrix} : b \in \mathbb{R} \right\}.$$

(c) For the group action in Exercise 2.3(a), for each value of $k = 0, 1, 2, 3, 4, 5$, we have

$$\text{Stab}(k) = \{g \in \mathbb{Z} : (2g + k)\,(\text{mod } 6) = k\}$$
$$= \{g \in \mathbb{Z} : 2g + k \equiv k\ (\text{mod } 6)\}$$
$$= \{g \in \mathbb{Z} : 2g \equiv 0\ (\text{mod } 6)\}$$
$$= \{\dots, -6, -3, 0, 3, 6, 9, \dots\}$$
$$= 3\mathbb{Z}.$$

Thus the stabiliser of each of the six elements of X is the subgroup $3\mathbb{Z}$.

(d)

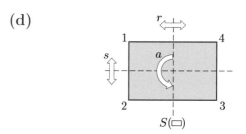

For the group action in Exercise 2.3(b), each vertex is stabilised only by the identity, so

$$\text{Stab}(1) = \text{Stab}(2) = \text{Stab}(3) = \text{Stab}(4) = \{e\}.$$

2.5

Subset A	Orb(A)	Stab(A)	|Orb(A)|	|Stab(A)|
◰	{◰,◳,◲,◱}	$\{e, b\}$	4	2
◩	{◩,◪,◧,◨, ◫,◲,◬,◭}	$\{e\}$	8	1
◫	{◫,◪}	$\{e, b, s, u\}$	2	4
⊙	{⊙}	$S(\square)$	1	8

In each case, the number of elements in the orbit multiplied by the order of the stabiliser is 8, the order of the group.

2.6 (a) The orbits are given by the columns of the action table. They are as follows:

$$\{1, 3, 5, 7\} = \text{Orb}(1) = \text{Orb}(3) = \text{Orb}(5) = \text{Orb}(7)$$
$$\{2, 4\} = \text{Orb}(2) = \text{Orb}(4),$$
$$\{6\} = \text{Orb}(6).$$

(b) Stab$(1) = \{e, z\}$,
Stab$(2) = \{e, b, x, z\}$,
Stab$(3) = \{e, x\}$,
Stab$(4) = \{e, b, x, z\}$,
Stab$(5) = \{e, z\}$,
Stab$(6) = \{e, a, b, c, w, x, y, z\} = G$,
Stab$(7) = \{e, x\}$.

(c) From parts (a) and (b), we obtain the following table.

x	|Orb(x)|	|Stab(x)|	|Orb(x)| × |Stab(x)|
1	4	2	8
2	2	4	8
3	4	2	8
4	2	4	8
5	4	2	8
6	1	8	8
7	4	2	8

In each case,

$$|\text{Orb}(x)| \times |\text{Stab}(x)| = |G|.$$

2.7 The orbit of $\{1,2\}$ contains the following elements:

$$e \wedge \{1,2\} = \{1,2\},$$
$$(2\ 3) \wedge \{1,2\} = \{1,3\},$$
$$(2\ 4) \wedge \{1,2\} = \{1,4\},$$
$$(1\ 3) \wedge \{1,2\} = \{2,3\},$$
$$(1\ 4) \wedge \{1,2\} = \{2,4\},$$
$$(1\ 3)(2\ 4) \wedge \{1,2\} = \{3,4\}.$$

As these are the six elements of X, we conclude that $\mathrm{Orb}(\{1,2\}) = X$, so there is just one orbit.

For each $\{i,j\} \in X$,

$$\mathrm{Stab}\,(\{i,j\}) = \{g \in S_4 : g \wedge \{i,j\} = \{i,j\}\}$$
$$= \{g \in S_4 : \{g(i), g(j)\} = \{i,j\}\}.$$

So $\mathrm{Stab}(\{i,j\})$ consists of those elements $g \in S_4$ that either fix each of i and j (so $g(i) = i$ and $g(j) = j$), or interchange i and j (so $g(i) = j$ and $g(j) = i$). For example,

$$\mathrm{Stab}(\{1,2\}) = \{\underbrace{e,\ (3\ 4)}_{\substack{\text{These}\\\text{elements}\\\text{fix}\\\text{1 and 2.}}},\ \underbrace{(1\ 2),\ (1\ 2)(3\ 4)}_{\substack{\text{These}\\\text{elements}\\\text{interchange}\\\text{1 and 2.}}}\}.$$

Similarly,

$$\mathrm{Stab}(\{1,3\}) = \{e, (2\ 4), (1\ 3), (1\ 3)(2\ 4)\},$$
$$\mathrm{Stab}(\{1,4\}) = \{e, (2\ 3), (1\ 4), (1\ 4)(2\ 3)\},$$
$$\mathrm{Stab}(\{2,3\}) = \{e, (1\ 4), (2\ 3), (1\ 4)(2\ 3)\},$$
$$\mathrm{Stab}(\{2,4\}) = \{e, (1\ 3), (2\ 4), (1\ 3)(2\ 4)\},$$
$$\mathrm{Stab}(\{3,4\}) = \{e, (1\ 2), (3\ 4), (1\ 2)(3\ 4)\}.$$

2.8 For each $(x,y) \in \mathbb{R}^2$,

$$\mathrm{Stab}(x,y) = \{r \in \mathbb{R} : r \wedge (x,y) = (x,y)\}$$
$$= \{r \in \mathbb{R} : (x, y+r) = (x,y)\}$$
$$= \{r \in \mathbb{R} : r = 0\} = \{0\}.$$

So $\mathrm{Stab}(x,y)$ consists of the identity element alone.

For each $(x,y) \in \mathbb{R}^2$,

$$\mathrm{Orb}\,(x,y) = \{r \wedge (x,y) : r \in \mathbb{R}\}$$
$$= \{(x, y+r) : r \in \mathbb{R}\}.$$

As r takes all real values, so too does $y+r$. Thus $\mathrm{Orb}\,(x,y)$ consists of all the points in $\mathbb{R}^2$ whose first coordinate is x; that is, it is the line through the point (x,y) parallel to the y-axis. Thus the set of orbits is the set of lines parallel to the y-axis. (This set partitions the plane, as expected.) See the following diagram.

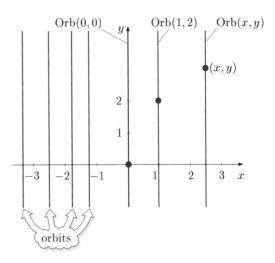

2.9 (a) First we find the orbit of any element of X, say $\{1,2\}$. We have

$$\mathrm{Orb}(\{1,2\}) = \{g \wedge \{1,2\} : g \in S(\square)\}.$$

We calculate the ten values of $g \wedge \{1,2\}$:

$$e \wedge \{1,2\} = \{1,2\},$$
$$(2\ 5)(3\ 4) \wedge \{1,2\} = \{1,5\},$$
$$(1\ 3)(4\ 5) \wedge \{1,2\} = \{2,3\},$$
$$(1\ 5)(2\ 4) \wedge \{1,2\} = \{4,5\},$$
$$(1\ 2)(3\ 5) \wedge \{1,2\} = \{1,2\},$$
$$(1\ 4)(2\ 3) \wedge \{1,2\} = \{3,4\},$$
$$(1\ 2\ 3\ 4\ 5) \wedge \{1,2\} = \{2,3\},$$
$$(1\ 3\ 5\ 2\ 4) \wedge \{1,2\} = \{3,4\},$$
$$(1\ 4\ 2\ 5\ 3) \wedge \{1,2\} = \{4,5\},$$
$$(1\ 5\ 4\ 3\ 2) \wedge \{1,2\} = \{1,5\}.$$

So

$$\mathrm{Orb}(\{1,2\}) = \{\{1,2\}, \{2,3\}, \{3,4\}, \{4,5\}, \{1,5\}\}.$$

Next we choose an element not in the orbit already found, say $\{1,3\}$, and find its orbit. We have:

$$e \wedge \{1,3\} = \{1,3\},$$
$$(1\ 2\ 3\ 4\ 5) \wedge \{1,3\} = \{2,4\},$$
$$(1\ 3\ 5\ 2\ 4) \wedge \{1,3\} = \{3,5\},$$
$$(1\ 4\ 2\ 5\ 3) \wedge \{1,3\} = \{1,4\},$$
$$(1\ 5\ 4\ 3\ 2) \wedge \{1,3\} = \{2,5\}.$$

(We do not need to check the other five values of $g \wedge \{1,3\}$, as all the elements of X have now been accounted for.)

So

$$\mathrm{Orb}(\{1,3\}) = \{\{1,3\}, \{2,4\}, \{3,5\}, \{1,4\}, \{2,5\}\}.$$

Thus there are two orbits, each with 5 elements.

(b) We have

Stab($\{1,2\}$)

$= \{g \in S(\bigcirc) : g \wedge \{1,2\} = \{1,2\}\}$

$= \{g \in S(\bigcirc) : \{g(1), g(2)\} = \{1,2\}\}$

$= \{g \in S(\bigcirc) : g$ fixes or interchanges 1 and 2$\}$.

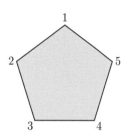

The only element of $S(\bigcirc)$ that fixes both 1 and 2 is the identity element, and the only element that interchanges 1 and 2 is (1 2)(3 5). Hence

Stab($\{1,2\}$) = $\{e, (1\ 2)(3\ 5)\}$.

Similarly, Stab($\{1,3\}$) consists of those elements of $S(\bigcirc)$ that either fix 1 and 3 or interchange them:

Stab($\{1,3\}$) = $\{e, (1\ 3)(4\ 5)\}$.

3.1 (a) The four vertices form a single orbit, so each orbit size is 4.

Each vertex is stabilised by exactly two symmetries in $S(\square)$, namely the identity and the reflection in the line of symmetry through that vertex. For example, vertex 1 is stabilised by e and s.

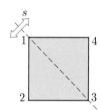

Therefore each stabiliser size is 2.

So, for each vertex x,

$|\text{Orb}(x)| \times |\text{Stab}(x)| = 4 \times 2 = 8 = |S(\square)|$.

(b) There are two orbits, $\{R, T\}$ and $\{S, U\}$, so each orbit size is 2.

Each line of symmetry is stabilised by exactly four symmetries in $S(\square)$, namely the identity element, the rotation through π, the reflection in the line itself and the reflection in the perpendicular line. For example, the line R is stabilised by e, b, r and t.

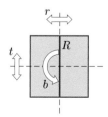

Therefore each stabiliser size is 4.

So, for each line of symmetry x,

$|\text{Orb}(x)| \times |\text{Stab}(x)| = 2 \times 4 = 8 = |S(\square)|$.

(c) There is just one orbit, with size 3, because each line of symmetry can be sent to each of the other lines of symmetry by a rotation of the triangle.

Each line of symmetry is stabilised by exactly two symmetries in $S(\triangle)$, namely the identity element and the reflection in the line of symmetry itself. For example, the line R is stabilised by e and r.

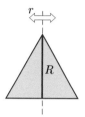

Therefore each stabiliser has size 2.

So, for each line of symmetry x,

$|\text{Orb}(x)| \times |\text{Stab}(x)| = 3 \times 2 = 6 = |S(\triangle)|$.

3.2 (a) Stab(2) = $\{e, (1\ 3)\}$.

(b) The left cosets of Stab(2) are

Stab(2) = $\{e, (1\ 3)\}$,

$(1\ 2)\,\text{Stab}(2) = \{(1\ 2) \circ e,\ (1\ 2) \circ (1\ 3)\}$
$= \{(1\ 2), (1\ 3\ 2)\}$,

$(2\ 3)\,\text{Stab}(2) = \{(2\ 3) \circ e,\ (2\ 3) \circ (1\ 3)\}$
$= \{(2\ 3), (1\ 2\ 3)\}$.

(c) By definition, each element of the coset Stab(2) sends 2 to 2.

For the coset $(1\ 2)\,\text{Stab}(2)$, we have

$(1\ 2) \wedge 2 = 1, \quad (1\ 3\ 2) \wedge 2 = 1$.

For the coset $(2\ 3)\,\text{Stab}(2)$, we have

$(2\ 3) \wedge 2 = 3, \quad (1\ 2\ 3) \wedge 2 = 3$.

So group elements in the same left coset of Stab(2) send 2 to the same element of X, and group elements in different left cosets send 2 to different elements of X.

3.3 (a) From the action table, we see that

Stab(4) = $\{e, u\}$.

The left cosets of Stab(4) are

Stab(4) = $\{e, u\}$,

$a\,\text{Stab}(4) = \{a \circ e, a \circ u\} = \{a, r\}$,

$b\,\text{Stab}(4) = \{b \circ e,\ b \circ u\} = \{b, s\}$

$c\,\text{Stab}(4) = \{c \circ e,\ c \circ u\} = \{c, t\}$.

(b) The set of elements that send 4 to 1 is $\{a, r\}$;

the set of elements that send 4 to 2 is $\{b, s\}$;

the set of elements that send 4 to 3 is $\{c, t\}$;

the set of elements that send 4 to 4 is $\{e, u\}$.

This is the same partition as in part (a).

3.4 From the solution to Exercise 3.3, the correspondence is as follows.

Left cosets of Stab(4)		Elements of Orb(4)
$\{e, u\}$	$\longleftrightarrow$	4
$\{a, r\}$	$\longleftrightarrow$	1
$\{b, s\}$	$\longleftrightarrow$	2
$\{c, t\}$	$\longleftrightarrow$	3

3.5 (a)

 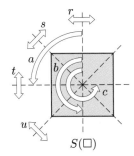

From the solution to Exercise 3.1(b), we have $\mathrm{Orb}(U) = \{U, S\}$ and $\mathrm{Stab}(U) = \{e, b, s, u\}$.

Each element of $\mathrm{Stab}(U)$ sends U to U. Each other element of $S(\square)$ sends U to S. Hence the correspondence is as follows.

Left cosets of Stab(U)		Elements of Orb(U)
$\{e, b, s, u\}$	$\longleftrightarrow$	U
$\{a, c, r, t\}$	$\longleftrightarrow$	S

(Theorem 3.2 tells us that the sets on the left *are* the left cosets of Stab(U). This is already clear for this example because Stab(U) has index 2 in $S(\square)$.)

(b)

 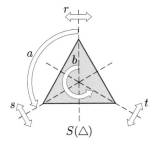

From the solution to Exercise 3.1(c), we have $\mathrm{Orb}(T) = \{R, S, T\}$ and $\mathrm{Stab}(T) = \{e, t\}$.

Each element of $\mathrm{Stab}(T) = \{e, t\}$ sends T to T. Also, the group elements a and s send T to R, and the group elements b and r send T to S. Hence the correspondence is as follows.

Left cosets of Stab(T)		Elements of Orb(T)
$\{e, t\}$	$\longleftrightarrow$	T
$\{a, s\}$	$\longleftrightarrow$	R
$\{b, r\}$	$\longleftrightarrow$	S

(Theorem 3.2 tells us that the sets on the left *are* the left cosets of Stab(T). You may like to check this by working them out in the usual way.)

3.6 (a) This is a group action. We show that axioms GA1, GA2 and GA3 hold.

GA2   For all $g, x \in G$,
$$g \wedge x = gx \in G,$$
by closure in G, so axiom GA1 holds.

GA2   For each $x \in G$,
$$e \wedge x = ex = x,$$
so axiom GA2 holds.

GA3   For all $g_1, g_2, x \in G$,
$$g_1 \wedge (g_2 \wedge x) = g_1 \wedge g_2 x = g_1(g_2 x) \qquad \text{(S.10)}$$
and
$$(g_1 g_2) \wedge x = (g_1 g_2)x. \qquad \text{(S.11)}$$
Expressions (S.10) and (S.11) are the same, by associativity in G, so axiom GA3 holds.

Hence $\wedge$ satisfies axioms GA1, GA2 and GA3, so it is a group action.

Remark   Some texts refer to this group action as the *left regular action* of a group. Although we did not use the terminology of group actions, our proof of Cayley's Theorem (Unit GTA3, Section 5) amounts to showing that a finite group is isomorphic to the group of permutations obtained from the left regular action.

(b) This is not a group action, since axiom GA3 does not hold. If $g_1, g_2, x \in G$, then
$$g_1 \wedge (g_2 \wedge x) = g_1 \wedge (xg_2) = (xg_2)g_1$$
but
$$(g_1 g_2) \wedge x = x(g_1 g_2).$$
These two expressions are equal when $g_1 g_2 = g_2 g_1$. This is not true in general, but does hold when the group G is Abelian.

As a counter-example, take G to be the group $S(\square)$, and let g_1, g_2 and x be the elements a, r and e, respectively. Then
$$a \wedge (r \wedge e) = a \wedge (e \circ r) = a \wedge r = r \circ a = u$$
but
$$(a \circ r) \wedge e = s \wedge e = e \circ s = s.$$

(c) This is a group action. We show that axioms GA1, GA2 and GA3 hold.

GA1   For all $g, x \in G$,
$$g \wedge x = xg^{-1} \in G,$$
so axiom GA1 holds.

GA2   For each $x \in G$,
$$e \wedge x = xe^{-1} = xe = x,$$
so axiom GA2 holds.

GA3 For all g_1, g_2, $x \in G$,
$$g_1 \wedge (g_2 \wedge x) = g_1 \wedge (xg_2^{-1}) = (xg_2^{-1})g_1^{-1}$$
and
$$(g_1 g_2) \wedge x = x(g_1 g_2)^{-1} = x(g_2^{-1} g_1^{-1}).$$
These expressions are the same, by associativity in G, so axiom GA3 holds.

Hence $\wedge$ satisfies axioms GA1, GA2 and GA3, so it is a group action.

Remark Some texts refer to this group action as the *right regular action*.

3.7 We show that axioms GA1, GA2 and GA3 hold.

GA1 For each $h \in H$ and each $g \in G$,
$$h \wedge g = hg \in G,$$
by closure in G, so axiom GA1 holds.

GA2 For each $g \in G$,
$$e_H \wedge g = e_H g = e_G g = g,$$
since the identity element of H is the same as that of G. Thus axiom GA2 holds.

GA3 For all h_1, $h_2 \in H$ and all $g \in G$,
$$h_1 \wedge (h_2 \circ g) = h_1 \wedge (h_2 g) = h_1(h_2 g)$$
and
$$(h_1 h_2) \wedge g = (h_1 h_2)g.$$
These expressions are the same, by associativity in G, so axiom GA3 holds.

Hence $\wedge$ satisfies axioms GA1, GA2 and GA3, so it is a group action.

3.8 (a) The group $\mathbb{Z}_n$ is Abelian, so each conjugacy class contains just one element—and 1 divides n.

(b) The group $S(\triangle)$ has three conjugacy classes:

the identity element:	$\{e\}$;
rotations through $2\pi/3$ and $4\pi/3$:	$\{a, b\}$;
reflections:	$\{r, s, t\}$.

The number of elements in each conjugacy class (1, 2 or 3) divides 6, the order of $S(\triangle)$.

(c) The group $S(\square)$ has five conjugacy classes:

the identity element:	$\{e\}$;
rotations through $\pi/2$ and $3\pi/2$:	$\{a, c\}$;
rotation through π:	$\{b\}$;
reflections in vertical and horizontal axes:	$\{r, t\}$;
reflections in diagonals:	$\{s, u\}$.

The number of elements in each conjugacy class (1 or 2) divides 8, the order of $S(\square)$.

3.9 We show that axioms GA1, GA2 and GA3 hol

GA1 For all $g \in G$ and all $h \in H$,
$$\phi(g) \in H,$$
so
$$g \wedge h = \phi(g) * h \in H,$$
so axiom GA1 holds.

GA2 Since ϕ is a homomorphism,
$$\phi(e_G) = e_H.$$
Therefore, for each $h \in H$,
$$e_G \wedge h = \phi(e_G) * h$$
$$= e_H * h$$
$$= h,$$
so axiom GA2 holds.

GA3 For all g_1, $g_2 \in G$ and all $h \in H$,
$$g_1 \wedge (g_2 \wedge h)$$
$$= g_1 \wedge (\phi(g_2) * h) \quad \text{(definition of } \wedge)$$
$$= \phi(g_1) * (\phi(g_2) * h) \quad \text{(definition of } \wedge)$$
$$= (\phi(g_1) * \phi(g_2)) * h \quad \text{(associativity)}$$
$$= \phi(g_1 \circ g_2) * h \quad \text{(homomorphism proper}$$
$$= (g_1 \circ g_2) \wedge h \quad \text{(definition of } \wedge),$$
so axiom GA3 holds.

Hence $\wedge$ satisfies axioms GA1, GA2 and GA3, so it a group action.

3.10 The elements of X form a single orbit, so $|\mathrm{Orb}(x)| = 4$, for each $x \in X$.

For each $x \in X$, $\mathrm{Stab}(x)$ consists of the elements of S_4 that fix x. There are six such elements, corresponding to the six permutations of the other three symbols. For example,
$$\mathrm{Stab}(1) = \{e, (2\ 3), (2\ 4), (3\ 4), (2\ 3\ 4), (2\ 4\ 3)\}.$$
Hence $|\mathrm{Stab}(x)| = 6$, for each $x \in X$.

As $|S_4| = 24$, we have, for each $x \in X$,
$$|\mathrm{Orb}(x)| \times |\mathrm{Stab}(x)| = |G|.$$

3.11 Axiom GA3 fails. Let g_1, g_2, $x \in G$; then
$$g_1 \wedge (g_2 \wedge x) = g_1 \wedge (g_2^{-1} x g_2)$$
$$= g_1^{-1}(g_2^{-1} x g_2)g_1$$
$$= (g_1^{-1} g_2^{-1})x(g_2 g_1) \quad \text{(associativity)}$$
$$= (g_2 g_1)^{-1} x(g_2 g_1) \quad \text{(S.1}$$
and
$$(g_1 g_2) \wedge x = (g_1 g_2)^{-1} x(g_1 g_2). \quad \text{(S.1}$$
In general, expressions (S.12) and (S.13) are not equal (though they are equal if G is Abelian).

As a counter-example, take $G = S_4$, with $g_1 = (1\ 2)$ $g_2 = (1\ 3)$ and $x = (1\ 2\ 4)$. Then
$$g_1 g_2 = (1\ 3\ 2) \quad \text{and} \quad g_2 g_1 = (1\ 2\ 3),$$
so
$$(g_2 g_1)^{-1} x(g_2 g_1)$$
$$= (1\ 2\ 3)^{-1}(1\ 2\ 4)(1\ 2\ 3)$$
$$= (1\ 3\ 2)(1\ 2\ 4)(1\ 3\ 2)^{-1} = (1\ 4\ 3)$$

and
$$(g_1g_2)^{-1}x(g_1g_2)$$
$$= (1\ 3\ 2)^{-1}(1\ 2\ 4)(1\ 3\ 2)$$
$$= (1\ 2\ 3)(1\ 2\ 4)(1\ 2\ 3)^{-1} = (2\ 3\ 4).$$
Thus axiom GA3 does not hold.

3.12 (a) We show that axioms GA1, GA2 and GA3 hold.

GA1 Let $g \in G$ and $xH \in X$; then
$$g \wedge xH = (gx)H \in X,$$
so axiom GA1 holds.

GA2 For each left coset xH in X,
$$e \wedge xH = (ex)H = xH,$$
so axiom GA2 holds.

GA3 For all $g_1, g_2 \in G$ and all $xH \in X$,
$$g_1 \wedge (g_2 \wedge xH) = g_1 \wedge (g_2x)H$$
$$= (g_1g_2x)H$$

and
$$(g_1g_2) \wedge xH = (g_1g_2x)H.$$
Thus
$$g_1 \wedge (g_2 \wedge xH) = (g_1g_2) \wedge xH,$$
so axiom GA3 holds.

Hence $\wedge$ satisfies axioms GA1, GA2 and GA3, so it is a group action.

(b) We have
$$\mathrm{Orb}(H) = \{g \wedge H : g \in G\}$$
$$= \{gH : g \in G\} = X,$$
since $\{gH : g \in G\}$ is the set of all left cosets of H in G. Also
$$\mathrm{Stab}(H) = \{g \in G : g \wedge H = H\}$$
$$= \{g \in G : gH = H\} = H.$$

Remark Recall that $gH = H$ if and only if $g \in H$ (see Unit GTA4, Exercise 1.3).

(c) If G is finite, then applying the Orbit–Stabiliser Theorem to the solutions to part (b) gives
$$|X| \times |H| = |G|;$$
that is, the number of left cosets of H in G, times the order of H, equals the order of G. From this result we can immediately deduce Lagrange's Theorem, thus obtaining it in yet another way.

3.13 (a) Any face of the cube can be sent to the position of any other face by, for example, a rotation of the cube. Hence the six faces form a single orbit. If F is any face of the cube then, by the Orbit–Stabiliser Theorem (Theorem 3.1), $\mathrm{Stab}(F)$ has order $48/6 = 8$. That is, there are eight symmetries of the cube that send a given face to itself.

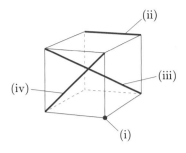

(b) (i) The eight vertices form a single orbit (since each vertex can be sent to each other vertex by a symmetry of the cube), so there are $48/8 = 6$ symmetries stabilising a given vertex.

(ii) The 12 edges form a single orbit (since each edge can be sent to each other edge), so there are $48/12 = 4$ symmetries stabilising a given edge.

(iii) There are 4 diagonals, and these form a single orbit (since each diagonal can be sent to any of the four diagonal positions), so there are $48/4 = 12$ symmetries stabilising a given diagonal.

(iv) Each of the six faces has two diagonals, making 12 face diagonals in all. These form a single orbit because any face diagonal can be sent to the position of any other face diagonal by a rotation of the cube. Thus there are $48/12 = 4$ symmetries stabilising a given face diagonal.

4.1 (a) By the Multiplication Rule, the number is
$$2 \times 2 \times 2 \times 2 \times 2 = 32.$$

(b) There are eight possible rings as shown below.

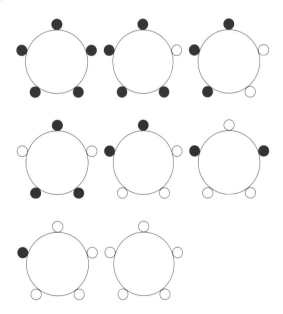

4.2 The sixteen coloured chessboards are drawn below; those that can be rotated into each other are placed together.

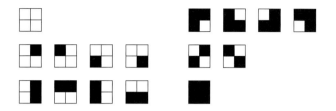

If we regard those that can be rotated into each other as the same, then there are six differently coloured chessboards.

4.3 For this group action,

Fix $(e) = \{R, S, T, U\}$,

Fix $(a) = \varnothing$,

Fix $(b) = \{R, S, T, U\}$,

Fix $(c) = \varnothing$,

Fix $(r) = \{R, T\}$,

Fix $(s) = \{S, U\}$,

Fix $(t) = \{R, T\}$,

Fix $(u) = \{S, U\}$.

4.4 The 10 elements of $S(\triangle)$ act on 2^5 coloured rings.

The identity fixes all 2^5 coloured rings.

Each of the four non-identity rotations fixes just two coloured rings—the single-colour rings.

Each of the five reflections fixes one bead and transposes the other beads in two pairs, so each fixes 2^3 coloured rings. (For a ring to be fixed by one of these reflections, each two beads transposed as a pair must be the same colour, so three beads—the fixed bead and one bead from each of the two pairs—can be coloured in any of the 2^3 possible ways and this determines the colours of the other two beads.)

By the Counting Theorem, the total number of orbits is

$$\tfrac{1}{10}(2^5 + (4 \times 2) + (5 \times 2^3)) = 8,$$

as illustrated in the solution to Exercise 4.1(b).

4.5 The four elements of $S^+(\square)$ act on 2^4 coloured chessboards.

The identity element fixes all 2^4 coloured chessboards.

The rotation through π fixes 2^2 coloured chessboards. (For a coloured chessboard to be fixed by this rotation, each square must be the same colour as its diagonally opposite square).

The rotations through $\pi/2$ and $3\pi/2$ each fix two coloured chessboards—the all-white one and the all-black one.

By the Counting Theorem, the total number of orbits is

$$\tfrac{1}{4}(2^4 + 2^2 + (2 \times 2)) = 6,$$

as illustrated in the solution to Exercise 4.2.

4.6 The four elements of $S^+(\square)$ act on 2^{16} coloured chessboards.

The identity fixes all 2^{16} coloured chessboards.

The rotation through π fixes 2^8 coloured chessboards. (For a coloured chessboard to be fixed by this rotation, each square must be the same colour as its diagonally opposite square. So the eight squares in one half of the board can be coloured in any of the possible ways and this determines the colours of the other eight squares.)

The rotations through $\pi/2$ and $3\pi/2$ each fix 2^4 coloured chessboards. (For a coloured chessboard to be fixed by one of these rotations, each square must be the same colour as the three squares onto which maps under successive quarter turns. So the four squares in one quarter of the board can be coloured in any of the 2^4 possible ways and this determines the colours of the other squares.)

By the Counting Theorem, the total number of orbits is

$$\tfrac{1}{4}(2^{16} + 2^8 + (2 \times 2^4)) = 16\,456.$$

Thus there are $16\,456$ coloured chessboards.

4.7 To determine the number of coloured cubes when only two colours are available, we rework the calculation in the solution to Question 3 (pages 41–42), changing the number of colours from to 2 wherever it occurs. This gives the answer

$$\tfrac{1}{24}\left(2^6 + (3 \times 2^4) + (6 \times 2^3) + (8 \times 2^2) + (6 \times 2^3)\right)$$
$$= 10.$$

So there are only 10 differently coloured cubes if on two colours are available.

4.8 (a) The 4 elements of $S(\square)$ act on 2^4 coloured blankets.

The identity fixes all 2^4 coloured blankets.

The rotation through π fixes 2^2 coloured blankets. (For a coloured blanket to be fixed by this rotation, each quadrant must be the same colour as its diagonally opposite quadrant. So two adjacent quadrants can be coloured in any of the 2^2 possible ways and this determines the colours of the other two quadrants.)

The two reflections each fix 2^2 coloured blankets. (For a coloured blanket to be fixed by one of these reflections, each quadrant must be the same colour the quadrant to which it is reflected. So two diagonally opposite quadrants can be coloured in an of the 2^2 possible ways and this determines the colours of the other two quadrants.)

By the Counting Theorem, the total number of orbits is

$\frac{1}{4}(2^4 + 2^2 + (2 \times 2^2)) = 7.$

Thus there are 7 differently coloured blankets.

(b) The seven possibilities are shown below.

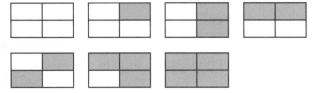

1.9 (a) The 6 elements of $S(\triangle)$ act on 2^4 coloured stickers.

The identity fixes all 2^4 coloured stickers.

The rotations through $2\pi/3$ and $4\pi/3$ each fix 2^2 coloured stickers. (For a coloured sticker to be fixed by one of these rotations, the three outer triangles must all be the same colour. So two triangles—the central triangle and one of the three outer triangles—can be coloured in any of the 2^2 possible ways and this determines the colours of the other two outer triangles.)

The three reflections each fix 2^3 coloured stickers. For a coloured sticker to be fixed by one of these reflections, each triangle must be the same colour as the triangle to which it is reflected. So three triangles—the outer triangle on the axis of reflection, the central triangle and one of the other two outer triangles—can be coloured in any of the 2^3 possible ways and this determines the colour of the remaining triangle.)

By the Counting Theorem, the total number of orbits is

$\frac{1}{6}(2^4 + (2 \times 2^2) + (3 \times 2^3)) = 8.$

Thus there are 8 differently coloured window stickers.

(b) The eight possibilities are shown below.

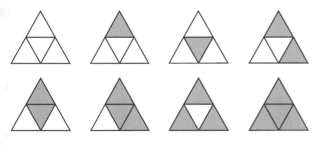

(c) To determine the number of stickers when three colours can be used, we rework the calculation in part (a), changing the number of colours from 2 to 3 wherever it occurs. The total number of orbits is

$\frac{1}{6}(3^4 + (2 \times 3^2) + (3 \times 3^3)) = 30.$

Thus there are 30 differently coloured window stickers in this case.

Index